THE SUPREME LAW

THE SUPREME LAW

By
MAURICE MAETERLINCK

Translated by
K. S. SHELVANKAR, M.A., Ph.D.

KENNIKAT PRESS/PORT WASHINGTON, N. Y.

521
M186

THE SUPREME LAW

Copyright 1935 by E.P. Dutton & Co., Inc.
Copyright renewal ©, 1963 by E.P. Dutton & Co., Inc.
This edition published by arrangement with E.P. Dutton & Co., Inc.
by Kennikat Press in 1969
Library of Congress Catalog Card No: 75-86042
SBN 8046-0571-8

Manufactured by Taylor Publishing Company   Dallas, Texas

ESSAY AND GENERAL LITERATURE INDEX REPRINT SERIES

CONTENTS

THE SUPREME LAW

I

NEWTON

IN the olden days they were not aware of gravitation or universal attraction. Neither the scientists and sages of India who pursued knowledge to the ultimate confines of the mind and possessed perhaps some secrets of a vanished world, nor the great astronomers of Chaldea and Egypt, ever asked by what miracle it was that the stars suspended in the sky did not drop pell-mell on to the earth which they thought was the centre of the universe, or go astray in space; or why men's feet which in dream seemed light as the feet of their gods were bound to earth by invisible chains.

Even the Greeks who imagined they knew

9

everything never realised what an enormous load pressed down on their dances and games, nor suspected the existence of the law of laws, beside which all others are but accessory rules and inconsistent caprices.

Nevertheless, about a hundred years after Jesus Christ, Plutarch glimpsed for a moment by a miraculous intuition the twin causes, mutually hostile, which hold up the moon in the sky. "The moon," he says in Amyot's delightful translation, "does not change according to the change in its weight, being held upright and prevented from slanting by the violence of its rotation." That is, in brief, the whole theory of gravitation.

Rather a strange man, by the way, this Plutarch, who had been initiated into all the secret sciences of antiquity and was the father of a discreetly romanticised form of history. It is to be presumed that his intuition was only a distant reflection of his hidden knowledge.

No attention was paid to it, and the mystery slumbered and was not disturbed until the sixteenth century of our era when Copernicus proved that weight is a sort of appetite (*quedam appetentia*, he said) which leads matter to aggregate spherically, and added that the influence of every celestial body must assuredly be felt on every other.

Some years later, Johannes Kepler brought forward three basic laws : the law of areas, the law of ellipses and the law of revolutions. In the law of ellipses he explains how—as Léon Sagnet has admirably summarised it—two neighbouring objects, situated alike outside the field of attraction of a third, attract each other in direct ratio to their mass ; and how the moon and the earth would crash into each other if they were not held in their orbits by some vital force.

He also proved that the movement of a body is naturally rectilinear and is deflected only under the influence of an external cause. His third law defines the ratio of the square of the revolution to the cube of the distance. It is obvious that he was hard on the scent, but he had no luck. He is indisputably one of the master minds of humanity, but a confused, visionary and above all unhappy genius who constantly mixed up astrology with astronomy. He went through a protracted and miserable inner struggle, and was obliged to earn his living by selling horoscopes and drawing up popular almanacks as Nostradamus had done.

In his *Astronomia Nova*, Kepler got very near to the heart of the matter and had a glimpse of gravitation, but failed to forestall Newton ; for, brushing past the great law, he missed it through not conceiving gravitation as passing through the centre of the earth, and being thus unable to demonstrate it he went astray. The prince of astronomers, of whom Newton is the king, seems to have been born under the most maleficent of stars.

At last Newton arose, twelve years after Kepler's death, years which, in Shakespeare's phrase, " bodied forth " the greatest event in history. It had been, as we have seen, preparing for some time and was in the air.

Let us recall in a few sentences how the law came into existence which unmasked the secret of the universe.

The famous anecdote related by Voltaire, who had it from Newton's favourite niece, Katherine Burton, tells us that the immortal mathematician of Woolsthorpe inferred from the fall of an apple in his orchard the law of gravitation.

But whence came to him the idea, which is much more important, of treating this attraction as determined in inverse ratio to the square of the distance ?

Newton owed much to Kepler who had blazed the trail for him. Making his calculations in accordance with the laws of the great German astronomer, he had already proved that the force of the sun would operate on the planets in proportion to the square of the distance. Wishing to find out whether the sun's action on the moon would confirm this law, he began his researches in 1666. Unfortunately, the prevailing estimate of the earth's circumference was not exact at that date.

Newton's calculations did not fit in with the facts. He therefore believed that the law was untrue, abandoned his investigations and did not resume them until sixteen years later the French astronomer, Jean Picard, had made a more precise measurement of the earth. As the truth emerged from these new figures Newton, choking with emotion and crushed by the weight of the immense veil he had lifted, had to request one of his collaborators to conclude his calculations for him. This time the hypothesis and the calculations concurred miraculously, the key to the universe was found, the capital law definitely established and applied to the billions of stars which people space.

.   .   .   .   .

Tantæ molis erat. . . . And one wonders whether there may not be awaiting us, in a future perhaps very near, some discovery equally enormous, equally simple and equally evident, to drag us with the same difficulties, the same gropings and delays out of our abysmal ignorance.

.   .   .   .   .

After the beginning of the eighteenth century the experiments conducted by Bouguer, Maskelyne, Airy, von Sterneck and others confirmed, through the deflection of a pendulum in the vicinity of mountains, the law of the attraction of masses. Subsequently the work of Cavendish, as resumed and continued by Cornu, Baille and Boys, determined the degree of attraction between two artificial masses, and the difference between the weight of the same object at the base and at the summit of a tower, thus verifying not now in the sky but on the earth itself the exactitude of the fundamental law of the world.

In the last and as it were testamentary page of his *Philosophiæ Naturalis Principia Mathematica,* which has been justly called the " loftiest production of the human mind," Isaac Newton tells us :

Hactenus Phænomena cœlorum et maris nostri per vim gravitatis exposui, sed causam Gravitatis nondum assignavi. Oritur utique hæc vis a causa aliquid quæ penetrat ad usque centra Solis et Planetarum, sine virtutis diminutione ; quæque agit non pro quantitate superficierum particularum in quas agit, (ut solent causæ mechanicæ) sed pro quantitate materiæ solidæ ; et cujus actio in immensas distantias undique extenditur, decrescendo semper in duplicata ratione distantiarum.

Rationem vero harum Gravitatis proprietatum ex Phænomenis nondum potui deducere, et Hypotheses non fingo.

Quidquid enim ex Phænomenis non deducitur, Hypothesis vocanda est et Hypotheses seu Metaphysicæ, seu Physicæ, seu Qualitatum occultarum, seu Mechanicæ, in Philosophia experimentali locum non habent.

It would be useless, I think, to translate this clear scientific Latin of. the seventeenth century. It was in truth the universal language for which we mistakenly wish to substitute the vulgar Volapuks or the barbarous Esperanto.

Let us note now the magnificent confession of ignorance made, at the very outset, by the discoverer of universal gravitation and all celestial mechanics, a man whose conclusions, as we shall see later, nothing has shaken or modified, not even Einstein's fantastic theories. Loftily and very heroically he declares that he knows nothing of the cause of the primordial phenomenon of the universe.

Let us add that the real discoverers of gravitation, Newton and Laplace, give us no clue as to what they think of it. The *Principia Mathematica*, with the exception of its last page, and the whole of Laplace's *La Mechanique Celeste*, constitute an inextricable jungle of geometrical figures and equations in which there is not a single clearing where a human voice might be heard. It was not until much later that Newton, questioned by his friends, agreed to give very briefly in letters not intended for publication some niggardly elucidation of the causes, the manner of action and, if one might use the phrase, the probable psychology of the law he had discovered.

No sooner were the proofs of the *Principia* corrected than claims were put forward in respect of the invention of the great law. One of his colleagues in the Royal Society, the

astronomer Robert Hook in particular, harassed
Newton with incessant and embittered protests.
It seems indeed certain that Hook had perceived
the law, but like Kepler had been unable to
demonstrate it. He had attempted to discover
if weight remained the same at different alti-
tudes. But his experiments, made on too small
a scale and with rudimentary instruments, had
not given appreciable results.

To have done with the business and for the
sake of peace, Newton agreed to declare in an
addendum to the fourth proposition that " the
inverse law of gravity holds good for all
celestial movements, and was alike discovered,
independently, by my countrymen Wren,
Hook and Halley."

Such coincidences, often fallacious and imaginary, but at times real, are quite frequent
in the great discoveries of the human race.
They are, as it is said, in the air. It may be
that some persons more sensitive than others,
or attentive, as receptors simultaneously catch
a hint from beyond the earth ; it may be that
the privilege is given them of extracting an

idea from the spiritual reservoirs that nature periodically throws open in time and space; or it may simply be that they are more lucky. At all events, the idea or the hypothesis is no more than a lovely corpse until it is realised, demonstrated, perfected and applied.

As another footnote it might be said that in the greatest discoveries the part played even by the greatest genius is always smaller than one thinks.

II

UNIVERSAL GRAVITATION AND CENTRIPETAL FORCE

LET us recall two or three elementary ideas which we must bear in mind.

The mass of a body is the quantity of matter to be found in it. Its volume is the quantity of space it occupies, and its density the mass contained in an unit volume.

The force of inertia, according to Laplace's definition, is the tendency of matter to maintain itself in its state of movement or repose.

Every object on earth, says Laplace, presses towards its centre in proportion to its mass, and so reacts on it and draws it towards itself with the same intensity.

The property which, again according to Laplace, the spheres possess of attracting as though their mass were concentrated at their centre, is very remarkable.

Finally, let us put *en exergue* the principle which sums up and presides over all that Newton has to teach us : Matter attracts matter in direct ratio to mass, and inversely to the square of the distance.

.   .   .   .   .

There is the great law, the essential law which governs the motions of all things.

It is often assumed that the Newtonian principles have been completely overturned by Einstein's theories. We shall see later that nothing of the sort has occurred. Einstein's theories only begin to be valid for bodies moving at rates of speed much greater than those of the planets.

" Classical mechanics," says J. Becquerel, " may be utilised where the square of the velocity of the bodies (velocity in relation to the observer) is concerned and ignored, perhaps, in respect of the square of the velocity of light."

Now the speed of stellar bodies is always, in comparison with that of light, very low. Beyond and below certain masses, certain distances and certain velocities, our calculations cease to be susceptible of being brought face to face with the reality, and degenerate very often into high-sounding hypotheses. Under all other conditions, in everything that is susceptible of verification, Newton's laws continue to govern the mechanics of celestial bodies.

Attraction, or the irresistible propensity of all particles of matter to cluster together, is the only universal law that suffers no exception and meets with no obstacle. A mass in Australia attracts a mass in London exactly as though the earth did not intervene.

" What distinguishes it from other physical phenomena," says Emile Borel, " is its immutability, its absolute independence of all external influences. Light is obstructed by opaque bodies, and deflected by a prism or a lens ; electrical and magnetic influences are modified by the proximity of certain bodies ; only gravitation remains invariably the same : there is no means of enhancing or diminishing it. It is indifferent alike to physical surroundings and the chemical nature of the bodies subjected to it. Radioactivity alone can furnish a similar example of an equally unchanging property, but while it is the property of a specific form of matter, gravitation is the same for all matter."

" We must nevertheless point out," adds M. Borel, " that an Italian scientist, M. Majorana, has quite recently obtained some results in regard to the absorption of gravitation by interposed bodies which, if confirmed, would be of very great importance ; but his experiments have so far been too few to merit discussion here."

. . . . .

I do not believe such absorption of gravitation by interposed bodies to be possible. If a body could absorb weight, what would happen to the weight ? Would it pass into the absorbant body like water into a sponge ? But then it would simply be displaced. If not, where would it go, and can it disperse without augmenting the weight of some thing? Or it may be that the interposed body acts as an insulator, withdrawing the object it protects from attraction? But would not the object thus insulated decompose and melt into space?

Anyhow, the experiments in question do not seem to have given satisfactory results, for we have heard nothing of them for ten or twelve years.

. . . . .

It is known that in a vacuum all bodies, whether light or heavy, fall at the same speed, and Eotvos's experiments have proved that the acceleration of weight at a given spot is the same for all bodies.

So long at any rate as we do not take into account rotation which, as we shall see later, is either derived from it or is its twin sister, gravitation is probably the sole source of all movement, and hence of all life on this earth in any case, and almost certainly everywhere else in the universe too.   There is no inquiry which attacks more directly and grips more closely the great mystery to which humanity seeks the key.

.   .   .   .   .

It is the law of laws, and manifests the will *par excellence*, the supreme will of the great All.

Its speed, as calculated by Laplace, on the basis of lunary perturbances, is a hundred million times that of light, that is to say, infinite.   The statement has, however, been contested by Cunningham, Heaviside and Eddington, but it seems to be corroborated by the velocity of some radiations.   The question is extremely complex and quite unreal, and James Mackaye, who devotes to it a long essay in his remarkable book, *The Dynamic Universe*, has no conclusions to offer us.

According to the most recent hypotheses gravitation, like light, is propagated by waves ; and the system of waves and the system of rays are the same.

" In the Newtonian theory, so splendid in

other respects, there is," says M. G. Darmois,
" a real gap. Nothing so gives us a sense of
inadequacy as the efforts made first by Laplace
and continued after him to introduce into the
theory an estimate of the velocity with which
gravitation is propagated."

It may be so ; but the wave theory, even as
revised and corrected by de Broglie who introduced into it the balistic or corpuscular conception, seems already questionable, and it is
very doubtful whether it applies to gravitation.

. . . . .

Could it be that the velocity of gravitation
is incalculable only because, being already
present everywhere and at the same moment,—
wherever one counts upon its activity, whether
it be to plunge a star into the abyss or hold a
pin poised—it has no need to displace itself
and hasten from the depths of unfathomable
space ?

. . . . .

It seems, in fact, that when there is a modification in any part of a material aggregation,

matter, which is everywhere and always cohesive, though scattered all over space, is instantly informed of the modification and reacts accordingly.

One could nevertheless imagine cases where it would not be present. Let us suppose that the earth, through the conjuncture of unforeseeable circumstances, becomes suddenly doubled in weight. Gravitational disturbances would occur all round in less than the thousandth of a second. But how long would they take to affect Sirius of which the light takes nine years to reach us?

·   ·   ·   ·   ·

We shall only begin to know what gravitation is when we begin to know what matter is. And of matter we know only so far some characteristics of probably minor significance. All that can be said in the meantime is that gravitation is the action of matter on matter.

·   ·   ·   ·   ·

Is it a spiritual force? It is impalpable, invisible, formless, without colour, without odour, without temperature, and silent as thought. Nothing can destroy or diminish it save itself, and its effects alone are felt by our muscles.

It emanates from, it is born of matter; but does not thought also emanate from matter, is not thought also born of it? As there is no thought without matter there is no attraction without matter. Is attraction the life of matter or of ether? It is of no consequence; it is itself life, or rather all life. If thought, as it seems to us, is the spiritual force *par excellence*, why should not gravitation be entitled to the same distinction? Does it not prove once more that matter and spirit are dual aspects, the one visible the other accidentally invisible, but both equally real, of the same being; and that all discussion of the subject is futile and childish?

.   .   .   .   .

Materialism, you say. All right; let us not be afraid of words. Is not a material spiritualism preferable to the spiritual materialism vaunted by so many innocent souls ?   It is in general those who have the least to do with spirit in any form who despise matter the most fervently and are the most convinced and uncompromising of " spiritualists."

.   .   .   .   .

How indeed do they come to be—universal gravitation and weight, its consequence—this curious law which is the soul of the universe ? We shall probably never know.   Anyhow, had it not existed from all time we should not be existing to-day.

.   .   .   .   .

Is there a limit to the power of attraction ? Since it manifests itself in inverse ratio to the square of the distance it follows that it must in the end be lost in infinitude or, at any rate, in the infinitesimal.   And there is some evidence for this ; but hundreds, even thousands of years would be needed to establish the fact conclusively.

It has been estimated, meanwhile, that the influence of solar attraction extends to a

hundred billion kilometres. Let us remember
that the sun is 92,833,000 miles away from the
earth. Flying at a speed of 187 miles an hour,
a pilot would take fifty years to reach it.
Neptune, the remotest planet in our system,
is four and a half billion kilometres from the
sun ; and our pilot flying at the same speed
would take fifteen centuries to get there.

If the sun's attraction extends to a hundred
billion kilometres, how far must the influence
reach of Antares, for example, the radius of
which is a hundred million kilometres greater
than the orbit of Mars ; or the influence of
the incredible Betelgeuse which can hold
twenty-eight million six hundred thousand
suns like our kingly sun, or that of Mira de la
Baleine which has a diameter of 124,274,000
miles ? It is true that, being still in a gaseous
state, their density is lesser than that of the
sun which is, too, itself in part gaseous.

.   .   .   .   .

When one considers that a star or a planet
is pulled in every direction, and itself pulls

from every point of its surface the orbs which surround it, it becomes impossible to visualise the inextricable, the bewildering tangle of forces that are in mutual conflict round each star in space; and one wonders how, torn in every direction, the wretched sphere yet continues to keep its balance. There are billions upon billions of stars which effect this miracle; else the universe would never have been other than an insane and all-pervading conflagration.

And what the stars accomplish on a stupendous scale, that the atoms in their turn accomplish in the infinitely small. For, from the time an atom comes into existence—and it has always existed—it presses down, it attracts and is attracted, it acts, it lives.

.   .   .   .   .

Is the universe falling in space? We know nothing about it. What we call falling is perhaps rising, turning round, deviating to the right or the left. In infinity there is neither height nor depth, neither ascent nor descent, nor any discernible directions. It may be that for all time we are being drawn directly, we and all the worlds we perceive, toward some unimaginable agglomeration of matter lost in the infinite of the infinite, of which it may be that we would only begin to have a glimpse millions or billions of centuries hence; and it may be that that agglomeration is itself neither a halting place nor a destination, but is hastening in its turn headlong towards an agglomeration more gigantic still, and so on indefinitely through limitless time and endless space.

.   .   .   .   .

If one could suppress gravitation which is like the power of God, interrupt or cut it short as an electric current is cut short, or neutralise it with centrifugal force, would it mean the end of all movement, i.e. of life, of which it is the unique source? Would it then be a form, almost comprehensible, of the nullity we seek in vain to conceive and which exists perhaps only in our imaginations?

.   .   .   .   .

It is meanwhile life itself, and matter, and acts, no matter where and in its smallest fragments, in proportion to the matter, that is to say the life, the force surrounding it.

And as no one has ever come across matter which had no attraction, which was, i.e. dead, we cannot imagine what it would be like. Would it continue to exist? We would not know, for we, too, should then be dead.

.   .   .   .   .

Were the stars motionless, time and space would cease to exist. Eternity alone would reign, the *Tota simul* of the Middle Ages, the moveless " Now " of which already in the sixth century Boetius had said : " The now which flees is time ; the now which remains immobile forms eternity."

As for space, rendered inert as time, it would become merely an invisible emptiness, resembling nullity ; if, indeed, the celestial bodies, while tearing it to shreds, do not re-knit it in the wake of their own destruction. It would no longer possess form, feature or justification. It would be everywhere and nowhere simultaneously, without change, without aim, hope or life.

.   .   .   .   .

Time and space are, however, but the two aspects of the same enigma. " There is no more intrinsic difference between time and length," Lindemann has said very justly, " than there is between length and breadth." One might say that time is the " space in flight," the *fliessende Raum* mentioned by Palagyi, as space is time congealed into eternity. Time, declares Alexander, is the mind of space, and space is the body of time.

.   .   .   .   .

It is said that there is no time external to the events of the universe, as it is maintained also that space exists only through the displacement or passage of matter. All that is true, but only relatively to us.

.   .   .   .   .

We have no senses with which to perceive space and time. We can neither see, hear nor feel time. And as for space, properly so called, we believe that we see it when we see an empty and limitless stretch before us, that is to say when we see nothing.

.   .   .   .   .

The ideal or pure space of our old geometry, traversed by no radiations whatever, has been called nullity. But space in this sense cannot exist; there are radiations the range of which has not yet been found to be restricted, and these would fill up such space on reaching it.

.   .   .   .   .

Gravitation creates time. And the intuition of man obscurely recognised it when he invented the first instruments for measuring the passage of days and nights: the hour-glass, the clepsydra, the pendulum clock and the sun-dial, the shadow on which is moved by the sun and the earth rotating and sinking in space. The sands that run down, the water that overflows, the weights that drop, all represent the cosmic force which guides the spheres in the infinite and enables childish man to measure with assurance that which does not exist—for time is but a delusion of creatures who cannot cope with eternity.

. . . . .

What would happen if time stopped? Nothing. We should have no suspicion that it had stopped. There *is* no time; there are only imaginary measurements of a thing existing only in our imaginations. When we seek to measure out with our small foot-rule a line that has neither beginning nor end, we undertake a childish and ridiculous task which means nothing and corresponds to nothing. It is the same when with our watches and almanacs we measure not time, which is but a phantom, but eternity, which is the sole reality.

Time can only stop when the rotation of the universe, or at any rate the rotation of all that is in the universe, stops. And whether it is a question of thousandths of seconds or thousands of light years it is always the rotation or the sidereal fall that is subdivided.

. . . . .

And if eternity came to a standstill? It always *is* at a standstill; always is stationary. It has no room in which to effect a displacement, and though to our eyes it seems packed with innumerable movements, none of them can step out of its infinity to occur elsewhere or go beyond it. Were it extinguished or did it vanish, holding everything within it as it does, nothing would be left behind. Not even space, for it is eternity which stretches and supports it. With its disappearance would disappear the spectacle of the universe as we see it to-day, to reappear, both together, in a different form.

. . . . .

Nothing can break away from the infinitude of eternity, for nothing has ever gone into it; and everything that does not exist and which apparently presents itself at the gateway would only represent an illusory attempt to re-enter on the part of that which had never come out.

.   .   .   .   .

Occasionally, at intervals of millenia, the spheres sink, or mount, and meet in the vast expanses of the sky other celestial bodies which attract them. Is *that* all the tragedy of space and eternity?

.   .   .   .   .

Centripetal force seems always to triumph on our earth. Everything culminates in the restfulness of its victory which we call the force of inertia. But it must be different in the other heavenly bodies; else the universe would have been congealed since its beginningless beginning and would now only be a limitless and motionless globe of inert matter.

.   .   .   .   .

The goal which all the spheres we observe desire to reach changes with a change in their position. But is it possible to imagine a goal in a time and in a space which have no limits ?   To have a goal is to go towards something ;   how can one go towards a thing when, at the same moment, one is everywhere ?

.   .   .   .   .

What would happen if gravity diminished by half ?   First of all, what would be the fate of our finest and most fragile organs, the brain, the heart, the lungs ?   Freed from an immemorial weight, the distracted heart, with its strength redoubled, would increase the violence of its beats and cause the aorta or the capillary vessels in the brain to burst.   And with air twice as light and thin the lungs would begin to pant as at the top of a mountain —which would probably mean a quick death for every one.

We cannot, however, assert anything definitely, for all experimentation is impossible ; if it is easy to rarefy the atmosphere, it never has been, it surely never would be possible to diminish the intensity of gravitation by the weight of a single hair—so long, at any rate, as one does not invoke the assistance of centrifugal force.

.   .   .   .   .

But let us suppose that these accidents do not occur. Our strength would then seem naturally twice as great, and like gods with winged feet, we should move with resilient and long undulating strides. Effortlessly we should jump over hedges and fences which we formerly considered insurmountable. We should be in the air as we are in water : with one spring we would be able to reach the fifth story of a building. Accomplished by muscles accustomed to the weight of the earth, all our gestures would acquire a disorderly quickness and amplitude.

Thinking to raise the arm to reach for a book, we shall find ourselves flinging it up violently towards the sky. With one bound we should be in our saddles, as though we had been hoisted up by an invisible crane and our horses would devour space like the legendary hippogryph. Our modest cars would do 250 to 350 miles an hour. Our children would shoot up rapidly and at maturity be twice as big as we are.

Mammals, fishes, birds, insects would all assume disconcerting habits and aspects. In the vegetable world, the sap no longer encountering any resistance, a sort of anarchic frenzy would be let loose. The growth of all trees and plants would be speeded up and attain dimensions difficult to assess.

. . . . .

But the marine phenomena—which would of course be awful—would with one stroke wipe out all these "anticipations." Feeling only a half of the formidable weight which chains it to its bed, the sea would overflow dykes and cliffs, in a tidal wave comparable to those of primeval times and regain possession of the globe. The annals of the earth would have come to an end, and the history of a new planet would commence in the depths of the waters.

. . . . .

If it is scarcely probable that the earth's gravity could diminish, it is almost certain, contrariwise, that the earth's weight and, consequently, its power of attraction, is constantly increasing. Every year, in the course of its transit through space, it seizes and seduces millions of bolides, aerolites, falling stars and millions of tons of cosmic matter, not to mention the asteroids strictly so called, i.e. small microscopic bodies,—the remains of a star which exploded or was pulverised—of which the biggest fragment, Vesta, has a diameter of 420 kilometres, and which represent, in the aggregate, a quarter of the earth's mass. A great many of these asteroids which revolve between Mars and Jupiter would in the end fall into our globe. Indeed the latter might well count, in apprehension or expectancy, upon many other sidereal acquisitions.

I leave it to those more capable than I am to imagine the phenomena which would occur in the wake of such augmentation of the earth's weight, and the spectacle of a humanity which, grown alarmingly heavy, would no

longer be able either to walk or to remain
erect and must crawl like worms. It is possible that the enhancement of weight would
become noticeable within two or three millenia.

.   .   .   .   .

It is quite possible, too, that this enhancement of weight has had, within the historical
period, a perceptible influence on the duration
of time ; at all events it *would* have had if it
were not counteracted by the contraction of
the earth's orbit which is tending, evidently,
to shrink closer to the sun.

It is also possible that the movement of
celestial bodies in our universe is continuously
accelerating, like the velocity of a falling
object. As there is nothing stationary around
us, we who must participate in that movement
would not be capable of measuring it. In any
case the increasingly rapid flight of the farthest
nebulæ towards the periphery of our bubble
universe is clearly proved by the displacement
of their spectral lines towards the red.

There is nothing to show that the rotation

of the earth on its axis and its revolution round
the sun have always had the same velocity and
range; there is, therefore, nothing to show
either that a day in the age of the Great Pyra-
mid, of Nineveh and of Babylon, was of the
same duration as ours. For all we know it
may have been twice or half as long, and no
one would have noticed the difference. We
have not, I repeat, any landmark, any fixed
standard, except our sidereal measurements,
and these are all moving with the same movement as ourselves.

.   .   .   .   .

If matter had been evenly spread through
space, in accordance with Newton's hypothesis,
how could we account for the fact that during
their interminable voyage across the infinite the
sun and the earth have not attracted and
absorbed this matter, and continued to grow
indefinitely at its cost?

And if matter had *not* been evenly distri-
buted, would there not have been differences
of pressure, and vacuums? And would they

not have been long since readjusted and filled up—these vacuums and uneven pressures ?

And would not all this lead to the formation of one enormous lump, the great danger or the final purpose of the universe ?

I do not believe it would. Being infinite, the world can never form a single block, for the block would be constantly increasing in size and always there would be something outside it to absorb ; and if one day it should cease to grow, for lack of nourishment, it would have become the universe which it had absorbed.

·　　·　　·　　·　　·

There exists perhaps a universe (or at any rate there exists *in* the universe) a world of pure thought. Most of the religious systems have taught us to hope that it exists, but we have had until now no example of thought without matter. Let us wait and see.

·　　·　　·　　·　　·

An electric current pressing through the wires round a bar of soft iron makes of the latter an *amant* more powerful than a natural lover. Is it not an analogous, or more spiritual, current which creates the force of attraction in matter ? And are we sure that this force never varies ? How should we know ?

.   .   .   .

Let us suppose that there is no gravitation. There would nevertheless continue to exist, in the endless fields of space and time, that which we call matter, represented not by atoms even, but by something more exiguous than we can conceive. Everything would be in a state of suspense, still, utterly motionless, eternal, and without hope ; a state that we should not be justified in naming as death, vacuum or nullity, for, as we know, it is not anything answering to these names. No word, of no human language, can adequately describe it.

We have, then, only to annihilate it for a moment in thought to recognise that attraction is the sole cause of all conceivable phenomena and the unique key to all the mysteries of the universe.

.   .   .   .   .

Without attraction, would not matter disintegrate to such a degree as to cease to be matter?   And, in *its* turn, could gravitation, when deprived of matter, i.e. when it must proceed from nothing to act upon nothing, continue to exist?

.   .   .   .   .

What would be a universe devoid of attraction?   It would be a universe without matter, which is, after all, quite possible, but would mean nothing to us; it would be a universe in which not only could we not remain alive but where we should be as inconceivable as it would be for us.   And in any case, even if we could exist there, we should not be able to appreciate it.

.   .   .   .   .

A god who wished to annihilate the world at one stroke would only have to deprive matter of its power of attraction.   Instantly everything would dissolve in what we can no longer call space, for, considering that only the movements and displacements of matter bring it into existence, there would no longer *be* any space.

   .   .   .   .   .

What would be the attractive power of an ounce of iron (if it were possible to preserve its cohesion) at a point in space beyond the reach of any attraction?

   .   .   .   .   .

By how many ounces are we the lighter because of the sun round which we revolve? Because of the moon? And of the great nebula towards which we are rushing? Were the earth solitary in space, remote from all gravitational influences, the attractive force of its centre would probably be such that we would no longer be able to remain erect. We owe our fragile lives, our shapes, our movements, all that we have and all that we are to the activity of all things that exist.

   .   .   .   .   .

It has been said that attraction or weight is a property as inseparable from matter as extension. It is even more so. One could at will enhance or reduce the extension or the volume of a bar of tin by rolling or compressing it; but its weight cannot be enhanced or reduced except by being transported to a different planet or up above the equator.

The attractive force of the earth, if not resisted by that of the more or less proximate stars, would represent exactly the weight of the globe. Professor C. V. Boys, after verifying in his laboratory the weight of a certain number of tons of lead, has calculated that the weight of the earth (in tons) is represented by the figure 6 followed by 21 zeros, a figure for which there is no designation, or no designation which conveys anything to the imagination, and which is written thus in astronomy: 6×10^{21}.

. . . . .

From the human point of view our muscles can by themselves appraise the earth's power of attraction by attempting, for example, to lift from the ground a dumb-bell of 150 or 160 pounds, that is of about our own weight. Is there some mysterious correlation between man's strength and the gravitational pull of the earth ?

.   .   .   .   .

Attraction affects small things as well as big. Bodies floating in equilibrium in a fluid which supports and prevents them from yielding to the call of the earth—which would cause them to drop to the bottom of the container—attract each other and, in the end, always cluster together.

.   .   .   .   .

To consider only our solar system. Born out of cosmic chaos, condensed in a nebula wherein arose a sun which threw off satellites, it is seeking its equilibrium through a certain stability which is constantly in danger and constantly in motion. But this stability is not proven ; and we possess in regard to it only some precarious assumptions which confirm its great antiquity. Besides, what could this stability be, if not repose, silence, immobility or death ; and why do we consider these more naturally as a goal than movement and life ?

.   .   .   .   .

Let us forecast the most probable manner in which our solar system would come to an end. The central star has absorbed in turn the nearest of the planets, Mercury, Venus, the earth and the moon. As its mass increases, so in proportion also its power of attraction ; and Mars, the giant Jupiter, Saturn and its rings, Uranus, and even Neptune, lost in the solitudes of space, are wrenched from their orbits, and merged into the sun.

What happens then ? Would these repeated collisions volatilise the system, and lead it back to the state of the original nebulæ, or would the mass, with its centripetal force multiplied, form one monstrous conglomeration ? Assuredly the phenomenon must have occurred more than once, in our universe or in others, in the course of the eternity prior to the moment at which we stand.

That being so, how comes it that centrifugal force, which must have run the greatest risks during these cataclysms, was not destroyed little by little to the advantage of the centripetal ? Must we believe, in spite of appearances, that they are, as we shall see later, both primordial and equivalent ?

.   .   .   .   .

Let us suppose that our globe has been pierced through from end to end, and that a tube or a sort of chimney or tunnel, cutting through the centre of the earth, links up the north pole with the south. Let us leave aside for the moment the hypothesis of a central flame, and questions of temperature and atmospheric pressure rendering life impossible below a certain depth.

Let us adapt to our purpose Eddington's celebrated hypothesis of a lift, and suppose that a sort of cabin or metallic chamber could slide, or rather fall, unimpeded, down the vertical tube or tunnel. Two or three investigators lock themselves up in this hermetically sealed chamber. From atop the north pole, where the cabin has been made fast at the mouth of the tunnel, it is released into the abyss.

To reach the centre of the globe, it must traverse a distance equal to the earth's radius, shall we say, exactly, measuring from the Pole, 840·581 miles. But in order to prolong the experiment and allow the investigators time to return, let us multiply the radius by ten, which would give us 39,494·388 miles, roughly a sixth of the distance from the earth to the moon.

No sooner than it is released, extremely curious phenomena would occur in the experimental chamber, as it descends at an ever accelerated speed. Let us ignore, for the moment, the physiological consequences which would probably be fatal. To begin with, the scientists who are locked up in the cabin would lose their weight and all sense of heaviness and lightness, for nothing around them could fall faster than themselves. Their apparatus and instruments would float freely in the cabin like bubbles of soap. If an object is released from their hands, it would remain poised in the emptiness.

Then, instead of accelerating furiously, as might be expected, beyond a certain point which mathematical calculations could easily determine with precision, the power of attraction would diminish in proportion as the cabin approached the centre; so much so that, arrived at the centre, instead of being carried away by its momentum and rushing on, through the second half of the tube, towards the south pole, it would gradually become immobilised there and cannot afterwards detach itself to pursue its southern course or re-ascend northwards. And, as during the descent, all sense of weight would have completely vanished.

Would the air in the cabin, ceasing to be held down, escape? And where could it go? By what would it be attracted? And would not the human body, no longer knit together by the centripetal force to which it has adapted itself in the course of thousands of years, lose its cohesion and disintegrate like a handful of ashes? Or rather, quite otherwise, would the cabin not have been crushed by the weight of a column of air 6000 kilometres high, before ever opposing forces began to neutralise the pressure?

And what would happen if, simultaneously, at each end of the tube, at the north pole and the

south, identical cabins were released ?   Would
there be an awful collision at the centre of the
earth ?   It is infinitely more probable that the
two antipodal cabins would come to a halt and
float peacefully in the central spaces, the occupants of each commiserating the fate of the
occupants of the other, imagining that they
would have to exist with their heads below and
their feet up in the air.

If, owing I know not to what discoveries, a
descent to the centre of the earth should one day
become possible, it would in all probability
reveal to us more cosmic secrets about gravitation, the heart of all mysteries, than a journey
to the moon.

.   .   .   .   .

But assuming that the air remains breathable
and under its normal pressure, would life be
yet possible in this cell where gravitation would
cease to operate ?   Would not our organs,
emancipated, and kept no longer in their places
by the exceedingly feeble attraction of our mass,
disintegrate or escape through every aperture
in our bodies ?   And that indefatigable pump,

the heart, meeting no longer with an enormous resistance to the ascent of the blood through the arteries and the countless capillary vessels, would it not beat frenziedly and fly to pieces like a " racing " engine, while the aorta cracks or the brain, fiercely flooded, brings to an end with a crashing hæmorrhage this drama of impossible evasion ?

It is possible, as M. Majorana hopes, that men would some day find out the means of interrupting, deflecting, transforming or annulling the telluric current of gravitation, but we would gain nothing from the most grandiose discovery mankind is capable of conceiving. We can no more live outside the zone of the earth's attraction than a fish out of water. Equally impossible would it be for us to subsist on the moon or any other planet or star which did not have approximately the same weight, that is to say, the same power of attraction, as our globe.

.   .   .   .   .

Before closing this chapter, let us return for a moment to the lift of which we have spoken ;

and suppose, instead of a fall straight down to the centre of the earth, an ascent upwards vertically, and with a velocity accelerating at the same rate as in the first instance. The phenomena of the earlier experiment would recur—in reverse order. The investigators' weight would increase in proportion to the increase in the speed of the ascent.

" Their length would merge into their breadth," as it says in *The Thousand and One Nights*, until there is a complete flattening out, or up to the moment when the cabin—or hollow bullet—is recovered by the earth or enters the zone of attraction of another star. In either case, the vertical ascent would probably change into a circular movement. It would be easy, but idle, to depict the disaster to which this impracticable experiment must lead.

.   .   .   .   .

As soon as we touch the question of infinity, in space or in time, we go out of our depths. There is nothing in common between the reality and what is imagined by our brains.

Everything takes place in a closed circle, rigorously watertight and sterilised. " The infinite is always imaginary," Joseph le Boucher has said admirably. It is imaginary though unimaginable; that is to say, our imagination both creates and annihilates it at the same moment. The infinite is impossible because it is human.

To the mind of a superman or an angel a triangle would still be a triangle; but his idea of infinity would have no relation with that which springs to our minds. Besides, the idea varies from man to man and has in common only a negation, more or less comprehensive, more or less intelligent, of the finite. Nevertheless we possess the certitude that infinity exists, for it is impossible to imagine that it does not, and its opposite, the finite, is even less admissible.

. . . . .

One of my correspondents, M. Raymond Delaunay, who has original and at times alluring notions in regard to gravitation, says that a body has weight only when put on another and bigger body. If it had no companions in space, its centre of gravity would be within itself and it would weigh on nothing.

He adds that two stars that resembled each other would not exercise mutual attraction. I believe that to be an error. Attraction, that nothing can curb, being in this case equal and reciprocal, would each augment or annul the other, according as there is an immediate or an intercepted contact.

.   .   .   .   .

When we speak of attraction, or gravitation, we necessarily speak of the universe, of the will of which it is the only unchanging manifestation. The universe has no movements other than those of gravitation and rotation. Within it, everything is in motion, but itself cannot change place, being omni-present. "If it is to displace itself," says M. Delaunay, "it must be subject to the influence of a centre of gravity situated outside of it, hence outside of space, which is impossible. The centre of gravity of the universe is *in* the universe."

" Did the universe turn round itself, the movement of rotation would correspond to immobility. A sphere cannot revolve or displace itself except in space. There is neither direction, nor height, nor depth in the void. The bottom of the universe is its centre." To which one might object that it *has* no centre— a centre always implies a limit. We shall meet with this hypothesis of rotation again in the chapter on the expanding universe.

.   .   .   .   .

III

UNIVERSAL ROTATION AND
CENTRIFUGAL FORCE

BY the side of centripetal force there arises another and antagonistic force, a born enemy—centrifugal force, which seems from the beginning (this is only a manner of speaking, for there is no beginning) to have prevented matter from forming into a lump capable of filling the cosmos.

What is this second force ? Is it, like good in evil, comprised in the centripetal force against which it is striving ? Is it the result of centripetal forces being brushed past by a body falling through infinity ? Is it only a centripetal force modified by another centripetal force ? Does it come from a primordial impulse which, like the centripetal impulse, has

61

had no origin ? Is it, like positive and negative electricity, the same energy at two different polarities (which is, of course, as in the case of electricity, too, only a play on words, elucidating nothing) ?

One thing is certain : like a god holding his own from the beginning of time against an equal, we see it battling—in the celestial bodies revealed by our most powerful telescopes as well as in the indivisible atom, which reproduces in the core of the infinitely small the movements of the most enormous stars—we see it battling uncompromisingly against gravitation.

．　　　．　　　．　　　．　　　．

If we do not know what centripetal force is in itself—being only acquainted with some of its effects—we are yet more ignorant of what centrifugal force is, the force that converts the vertical fall of a body into an ellipse. Newton called it the " transverse force," as opposed to centripetal forces. This transverse force, at a given moment, arrests the descent of planets towards the sun and compels them to trace an elliptical orbit maintained indefinitely in a space devoid of friction.

.   .   .   .   .   .

Jeans declares that, if we could shoot horizontally from the top of a mountain a bullet charged with a speed of about six and a half kilometres per second, instead of vanishing in a straight line into infinity, the bullet (assuming that Jeans has eliminated from his hypothesis the friction of the air) would fly round our globe until its initial energy is exhausted : the earth's gravitation would exactly neutralise the horizontal force.

It is in fact likely that the case of the moon is more or less analogous to that of the bullet ; and the hypothesis provides us with an ingenious image which shows in miniature what we see in the heavens without at all understanding it.

.   .   .   .   .   .

For the time being, and by way of summary, it may be said that some centrifugal forces are but centripetal forces in a more or less complicated and mutually counteracting state; while others, and in particular those of rotation—we shall come back to them later—are autonomous. All this, however, does not suffice to explain everything, indeed explains nothing; and we would probably have to look in the direction of ether, or if we do not admit the term, in the direction of whatever is admitted to take its place; that is to say, in the direction of space or certain properties of the universe of which we are still totally ignorant.

.   .   .   .   .

Here, in a letter to Bentley, the confidant of his thoughts, is what Newton said of gravitation, of which he had just discovered the laws :

" That Gravity should be innate, inherent and essential to Matter, so that one Body may act upon another at a distance through a Vacuum, without the mediation of anything else, by and through which their Action and Force may be conveyed from one to another, is to me so great an Absurdity, that I believe no Man who has in philosophical Matters a competent Faculty of thinking, can ever fall into it. Gravity must be caused by an Agent acting constantly according to certain laws; but whether this Agent be material or immaterial, I have left to the consideration of my Readers."

This is a confession of ignorance, and a delimitation of the subject, which the recent and noisy exploits of our relativists and astrophysicists have scarcely broken down.

As regards the transverse or centrifugal force, it is, according to Newton, inexplicable through gravitation. In his second letter to Bentley, he says actually: "I do not know any Power in Nature which would cause this transverse Motion without the divine Arm." And further on, in the same letter, he adds: "Gravity may put the planets into motion, but without the divine Power it could never put them into such a circulating motion as they have about the sun."

And elsewhere: " God has placed the fixed stars at an immense distance from one another, fearing lest the orbs should fall the one on the other through the force of gravitation."

. . . . .

There is the appeal to God, the tragic and final gesture of the perplexed scientist who does not wish to own that he has ceased to understand, and cloaks his ignorance with a word of which the meaning is still more obscure than the meaning of all other words in all the languages of man. There is the confession we encounter at the very origins of human thought, in the great religions of India.

What have we found out since?

"Something is doing something we do not know what," writes Eddington. Is not this *nescio quid*, which is the last word of our science, but a faint and vulgar echo of the magnificent avowal of the Sama Veda saying of the supreme Deity: "He who believes he knows it not, knows it; he who believes he knows it, knows it not at all. It is regarded as incomprehensible by those who know it most, and as perfectly known by those who are utterly ignorant of it."

A confession re-echoed three or four thousand years later in the Zohar where God becomes a mark of interrogation in the darkness; next in the writings of the Areopagite and of his disciple, Scotus Erigena, the great theologian of the ninth century, where God is the being without attributes, that is to say almost nullity, and the incomprehensible essence of the universe; and in the *Deus qui melius scitur nesciendo* of St. Thomas Aquinas and of Bossuet, that most rigid and orthodox of the theologians of the Catholic Church who asserts that "the whole vision of faith seems to be reduced to seeing well that one sees nothing."

.   .   .   .   .

According to the illustrious astro-physicist J. H. Jeans, the origin of universal rotation should be looked for in what he calls the " gravitational instability " of the huge mass of chaotic gas out of which all celestial bodies come. There are currents in the chaos which take, in the nebulæ formed therein, a circular movement. In all the worlds which have issued from these nebulæ, and until their final dissolution in space or their ultimate condensation into white dwarfs, this movement persists.

.   .   .   .   .

This explanation presupposes chaos. We describe as chaos all that took place before man arrived and diminished the world to the measure of his petty life. But chaos has never, could never have existed. The same laws which to-day regulate and order all things that exist were already governing their movements. They have not been formed little by little, these laws : they coexist with everything, from all eternity, and without them nothing would have come into being. They are life itself and form a part of all that they control.

Nevertheless, and provisionally, let us assume chaos. Jeans' explanation is admissible. However, like all explanations of the unexplainable, it merely shifts the position of the question. What are these currents ? Like the brownian movement in terrestrial fluids, have they, equally with the gaseous mass wherein they prevail, always existed ? If not, they were produced therein. Why ?

If they had always existed, one would suppose that the rotations, which derive from and are but transformations of them, would still continue ; for a transformation is not the end. Like gravitation, the rotations would be an essential form of the *primum movens* ; and it might be thought, from that, that the two have coexisted throughout eternity, being indispensable to the equilibrium of the universe as we see it—a universe which, deprived of them, would never have been other than a lump of matter drifting without aim or purpose in a space which would be strangely like a vacuum and a nullity, that is to say, in a space which would not be existent at all.

.   .   .   .   .

If, on the other hand, the currents originate in the gaseous mass, the rotations caused by it must of necessity—since they have had a beginning—come to a stop, and everything must some day terminate in one lump of matter.

However, in the one case as in the other, we have the same insoluble enigma of movement and life.

Such is the spectacle our astronomers see in the heavens. Everything which occurred at the beginning, that is to say throughout eternity, is still unfolding itself before our eyes; and nothing changes in that which is ceaselessly changing.

.   .   .   .   .

What are we to say of the periodical comets ? Of the others, millions and millions upon them, which circulate through the sky, we know nothing. Let us take, for example, Halley's comet which returns every seventy years, and the great comet of 1843 which will not be visible again until the year 3808. At its perihelion, Halley's comet is closer to the sun than Venus; and at the other end of its elliptical orbit, it goes beyond Neptune into the boundless trans-Neptunian wilderness.

As for the great comet of 1843, which has a tail 320 million kilometres long, that is to say more than twice the distance from the earth to the sun, on the 27th February, 1843, it passed 52,000 kilometres away from the sun—on which protuberances more than 200,000 kilometres in height have at times been observed—and then went off to wander in the ultra-planetary spaces for another 1950 years.

Now these magnificent stars, these incredible sidereal rockets, have but insignificant nuclei. That of the 1843 comet, in particular, has a diameter of only 900 kilometres; and this tiny nucleus, rubbing against the photosphere of the gigantic star at the centre of our universe, was not only not whisked off, but rushed along at a speed of 550 kilometres per second, to fly to the other end of its orbit which it would only reach after 975 years.

What has happened to centripetal force in all this? What unimaginable and mad power sets it aside during these manifestations, so prodigious, and transgressing all the laws we think we have discovered? At bottom the astronomers know absolutely nothing about it.

. . . . .

Meanwhile, are not things moving towards that state where one enormous lump of matter would fill the cosmos? Should we not be already entombed in that lump if, at some moment prior to its complete formation, it had not burst? But where does the force come from, what is it, which has caused or could cause it to burst? What name can we give it which would not, after all, suggest a form, an effect, an influence or more probably an impact of centripetal force?

Until we have penetrated the mystery of the equilibrium and eternity of these two cardinal forces, we shall know nothing. They do not so far appear to have been sufficiently studied, although of all the powers which surround us attraction is the most familiar one. It enters into every heart beat of ours, into every gesture that it assists or hinders, into every part of our lives of which it is the real source.

. . . . .

Our centrifugal force, the centrifugal force of our planet, which we can to some extent study and verify, arises out of the rotation of the globe. Nil at the poles, at the equator it makes a weight of three kilos lighter by ten grammes. The astronomers say that it has a force there equal to the 289th fraction of terrestrial attraction. In other words, if the rotation of the globe were seventeen times faster, there would be no weight at the equator. What would be the fate of a humanity which, by moving towards the tropics, could thus unburden itself of the uncanny weight which is crushing it down ?

. . . . .

But why does this rotation occur ? What has set it going ? Whether it be the result of the explosion of the solar nebula and the projection of planets into space ; or of a collision with a vagrant star ; or, as is more probable, of the passage of a forked nebula through the stuff of a cosmic cloud—in any case, already at the very beginning of the formation of our world we find a rotation which determines the majority of its phenomena. But‛ how has the rotation been able to maintain itself and resist for so‛ long the all-powerful centripetal force ? Would it not have been annihilated thousands or millions of centuries ago if it were not at bottom either an autonomous force, or a form, ceaselessly resurrected, and incomprehensible, of centripetal force ?

．　　　．　　　．　　　．　　　．

It is almost certain that the stellar movements we believe to be circular or elliptical are in reality spiral ; and that the existence of humanity has been all too brief to enable us to measure them. Is it unlikely that the moon would ultimately fall into the earth and the earth into the sun ? Meanwhile the moon does not drop down because it soon exhausts the energy of its initial propulsion in a straight line, which is quickly curved by the earth's attraction, and because, again, it is held in place by the centripetal force of the sun.

But what is the force that prevents the earth and the other planets from falling into the sun ?

Is it not the same feat here—the deflection of
a straight line that is subjected to centripetal
forces and gets broken up into ellipses and
spirals—a feat that the bullet concludes in a
few hours or days, but which the immensity of
a planet can only accomplish in some millions
of centuries, time being—do we not know it?
—but the slave of density?

.   .   .   .   .

But whence do these antagonistic forces
come? Must we say, as Newton said, that
it is the hand of God? Or as they say to-day,
that centrifugal force is only an interwoven
form of centripetal forces opposing one another? In which case Newton would rediscover the hand of God in the centripetal
forces, and we should only have taken a
deceptive step in the dark.

.   .   .   .   .

Is it necessary, however, to believe that
centrifugal force has issued from or is sub-
ordinate to centripetal force? Has it not been
established that rotation is as universal as

gravitation ? As you will not find a particle of matter without electrons, so you will not find a star in the sky which does not turn round itself and round something else. What would be a star, what would become of a star without rotation ? We cannot say, for we have no knowledge about it at all.

The moon, which for us is the archetype of a dead star, turns on its axis and round the earth with a regularity revealed to us by observations made over two thousand years, and by the oldest astronomical charts. And if we continue to see the same aspects of it, it is on account of the admirable, the faultless synchronisation of its movements with those of our globe—a synchronisation hardly disturbed by the libration, such as oscillations visible in some parts of it situated close to the rim of the dark half.

.   .   .   .   .

As soon as you admit centripetal force, on the same grounds and with as much justification, you ought to admit universal rotation, the effects of which are invariably centrifugal, and which can be verified in infinitely small objects as well as in those infinitely big. Like centripetal force, it is the very life of matter. We see less of it, because it operates mainly in the interior of the atom where the electrons revolve restlessly round the nucleus, and manifest an energy superior perhaps to that of gravitation. If a dead star were to hide itself somewhere in the wilderness of the skies, this internal energy would subsist in it and perhaps, ultimately, break out into the open.

IV

THE WHITE DWARFS

THE white dwarfs to which we have alluded in a previous note deserve special mention as they represent the final effort, the supreme triumph of attraction.

We confront here the strangest enigma of a world, the sidereal world, that swarms with mysteries. Owing to their extremely feeble luminosity, astronomers have been able to discern so far only four white dwarfs, including the satellite of Sirius and the star van Maanen. They are, however, in all probability numerous in space where they constitute what are called " dark clouds " composed of obscure matter which reduce or put out the glow of some stars, that among others of the nebula of Orion.

We must not be misled by the term dwarf, for they are not minute. It has been calculated, for instance, that the mass of Sirius's companion is eight-tenths that of the sun and that its density, in relation to that of the latter, is 37,800. Some astronomers claim that the atoms in these fabulous stars are destitute of electrons. I do not believe that matter can be found separated from electrons or that an electron can die. But it may be granted that, if the atoms were completely crushed in and had no room to move, they would remain inert until they were liberated.

In any case, the compression of matter there is as the compression of a ton of matter in a liqueur glass. It is estimated through other calculations that the density of the satellite of Sirius is two thousand times the density of platinum, our heaviest metal, and fifty thousand times the density of water. The density of the star van Maanen, which is smaller, is three hundred thousand times the density of water. In comparison with these monsters hundreds of times harder and more impenetrable than diamond, we are only, with our rocks and metals, inconsistent nebulæ, hazy fumes distorted or dissolved by the slightest breath out of space.

Why, after all, should we not be nebulæ? Everything is relative. And perhaps the true, the final condition of matter, that to which it seems to be tending is the condition of Sirius's satellite or the star van Maanen.

.　　.　　.　　.　　.

What was the force which availed to compress matter so much as to press out of it all the space it contained, as they squeeze oil out of the olive under the grinding stone? I have not found any satisfactory explanation of this in the writings of the astro-physicists.

.　　.　　.　　.　　.

Gravitation being proportional, as Newton said in his *Principia*, to the quantity of matter contained in a body, the attractive power of the white dwarfs should be fabulous. Is it not to that we must ascribe some sidereal disturbances yet unexplained? Let us add that at that degree of compression matter becomes incapable of expansion and cannot therefore get cold either. It constitutes such an accumulation of power that it seems to be definitively dead, exempt from all manner of evolution and annihilation, as though it had reached its culminating point, its eternal condition. One would say that nature had led it astray into a blind alley from which only a collision with a world of equal density could rescue it.

.   .   .   .   .

Are these white dwarfs either the earliest adumbrations or the last vestiges of a single mass which the universe was or which it is going to be? Is it the end or the beginning of everything? We have no means of knowing, says Jeans, whether the sun or a star has spent a part of its existence in the condition of a white dwarf.

Yet Jeans is inclined to believe that if the luminosity of our sun diminished by three per cent—according to its position on Russell's diagram—the star on which our fate depends, and the fate of all the planets, would rapidly contract and become a white dwarf similar to the satellite of Sirius. We may, however, be reassured. It hardly seems that this reduction of three per cent can take place for another fifty million years.

In fact the sun, as Russel Dugal Stewart, of the Princeton Observatory has said, is only a dwarf that has not yet lost all its splendour. It is a star but faintly luminous and become already yellow and very dense.

.   .   .   .   .

What would happen—though the hypothesis is unlikely on account of the density of the star—if a fragment of a white dwarf detached itself and fell into our globe? A piece of this diabolical matter, no bigger than a pin-head, would weigh more than two kilos and pierce through the hand like a revolver shot. One wonders why bodies of such weight do not fall into space, playing havoc with all the spheres that they must attract to themselves in the course of their passage. They fall, indeed, but do not strike anything; and as they are falling and always will fall, it is as though they did not fall at all.

· · · · ·

What would have happened if a vacuum could exist somewhere in space ? Would the matter scattered in its environs rush into the gap ? But is it not matter that attracts matter across a vacuum and not the vacuum itself which, being nothing, can attract nothing ?

V

ETHER

IT has been said very rightly that in antiquity they knew only of one substance :
matter ; and that the study of the phenomena
of light has led to the recognition of a second :
ether. Yet while air and even light have been
weighed, it has not so far been possible to
weigh ether, either because, as some say, it does
not exist, or because we have not the requisite
instruments ; or, it may be because, as is
maintained by those who consider it indispensable for the explanation of a certain number
of phenomena, the earth's attraction does not
act upon it.

Newton endeavoured to account for gravitation by invoking differences of pressure in
the ether ; but he did not wish to publish
his theory because he declared himself " unable

to give, from experiment and observation, a satisfactory account of the conditions and the manner in which it produces the capital phenomenon of nature."

But he readily answered questions put to him by his friends. Thus, in a letter to Robert Boyle, the scientist who discovered several years before Mariotte the law which bears the latter's name, he set forth his conception of ether:

" I shall set down one conjecture more, which came into my mind now as I was writing this letter: it is about the cause of gravity. For this end, I will suppose ether to consist of parts differing from one another in subtlety by indefinite degrees: that in the pores of bodies there is less of the grosser ether in proportion to the finer, than in open spaces; and consequently, that in the great body of the earth there is much less of the grosser ether in proportion to the finer, than in the regions of the air: and that yet the grosser ether in the air affects the upper regions of the earth, and the finer ether in the earth lower regions of the

air in such a manner that, from the top of the
air to the surface of the earth, and again from
the surface of the earth to the centre thereof,
the ether is insensibly finer and finer. Imagine
now, any body suspended in the air, or lying
on the earth ; and the ether being, by the hypothesis, grosser in the pores which are in the
upper parts of the body, than in those which
are in the lower part ; and that grosser ether,
being less apt to be lodged in those pores, than
the finer ether below ; it will endeavour to get
out, and give way to the finer ether below,
which cannot be, without the bodies descending
to make room above for it to go out into."

This is evidently not very clear, and flounders
about in the unknown, but it does not seem at
bottom to be more obscure than present-day
theories. However that may be, it is interesting
to quote the views in regard to gravitation of
the genius who discovered it, and who, like all
mortals, stammered and stumbled in face of the
enormous riddle.

Is centrifugal force caused by the resistance,
or friction, of ether when a body falls
vertically ? And is the vertical fall—it is
indeed very probably only an illusion—in its
turn only a movement of ether, similar to the
eternal and inexplicable brownian movement
in the infinitely small particles in a drop of water
which seems ascribable as some scientists,
notably Wiener, think, " only to an internal
movement characteristic of the state of fluid-
ity " ? That, of course, only shifts the problem
without solving it.

.   .   .   .   .

But has it been proved that ether exists ?
The famous Michelson-Morley experiments are
too well known to be usefully described here ;
their details may be found in every book on
Relativity. The first of these experiments dates
back to 1881. It was resumed in 1887 by
Michelson and Morley ; and with an increasingly rigorous precision, by Morley and Miller
in 1904–1905. The first therefore antedates
Einstein's theories by more than thirty years.

These experiments have shown that no displacement in relation to ether can be verified by utilising the movement of the earth round the sun. Ether has always been elusive, and it has not been possible to discern it even as a kinetic point of reference. To conclude from this that it does not exist at all is but a step, and the step has perhaps been taken too lightly.

First of all, as Lieutenant-Colonel Corps has very rightly pointed out, it must be said that " experiment has only been made with speeds which, however great they may be in relation to those observed on our planet, are yet so feeble in relation to that of light that they may be considered to be, and have in fact been so considered, in calculations as infinitely small quantities of which the superior powers might be neglected. It is therefore permissible to wonder if the negative result of this experiment would hold good for velocities approximating to that of light."

What, indeed, could be the effect of the 18·641 miles per second—which represents the speed of the earth round the sun—when it is introduced into calculations involving the 186,412 miles per second of light?

In the second place the experiment did not reveal, as was anticipated of it, the total movement of the earth with reference to the ether which fills space, because, say Lorentz and Fitzgerald, the movement of a body with reference to ether produces a contraction of that body in the direction of the movement, so that the contraction makes up for the lengthening of the trajectory of the light ray and thus confers on ether the existence which it is denied elsewhere.

Einstein says that the effect anticipated before the Michelson experiment was not produced because the absolute axes of Newtonian mechanics are inaccessible. Absolute space, or ether in repose, does not exist, for if its only function were to remain hidden it is a gratuitous supposition to assert that it exists.

Let it be said in passing that arguments of this sort could lead us far astray.

.   .   .   .   .

However, the hypothesis of ether is indispensable for the explanation of all phenomena of light, electricity, magnetism and probably gravitation. The inexistence of ether would involve the inexistence of light. "Without ether," says Lieutenant-Colonel Corps, "not only do the laws of the velocity of light become paradoxical, but the phenomenon of light is itself incomprehensible. We cannot conceive of it in fact save as a movement, either a direct movement as in the theory of emission or the propagation of a vibratory movement, as in the wave theory. Now vibration assumes something which vibrates and which cannot, without ceasing to exist, cease to vibrate or become immobile, and immobile in an absolute fashion, since vibration cannot simultaneously depend on the system from which it is emitted and on that by which it is received, and must consequently be independent alike of both."

.   .   .   .   .

In default of ether we shall either have something which looks like its twin brother or an absolute vacuum. But what is absolute vacuum? If there should be anything—and surely there is something—in the universe, the vacuum would be filled and exist no longer. No matter what it is, light, electricity, cosmic influence or gravitational force, if it went near the vacuum it would be absorbed by it, and having absorbed it the vacuum would cease to exist. It is the parent or the offspring of nullity, as impossible, as inconceivable as the latter, which, being nothing, can answer no questions.

If the vacuum in space were absolute it would suck in, absorb all the celestial bodies, whereas that which is actually in its place—in the place of the vacuum—presses on the celestial bodies like air round a soap bubble, and compels them to take a spherical shape.

It is, moreover, almost certain that the earth, in spite of its enormous centripetal force, would explode in this vacuum like a rubber bladder in the bell of a pneumatic machine, for the suction of the vacuum would prevail over the gravitation of the mass.

.   .   .   .   .

But when we speak of an absolute vacuum we are playing with words. Absolute vacuum is a metaphysical entity which we believe it possible to imagine but which is at bottom unimaginable. When we have done our best to produce an utter vacuum in a tube, a bulb or a bell of glass or steel we have only exhausted or driven the air out of it; but this air has made way for something that was already there and that we can neither pump out nor expel.

Besides, there is nothing to show that, as the expelled air cannot re-enter the tube or bulb, the ether—which knows no obstacle and penetrates everywhere since it is everything—replaces it and facilitates the electrical phenomena manifested in the relative vacuum that only begins to live its characteristic life when it is rid of the element, the too heavy element, suffocating it.

.   .   .   .   .

Nullity, or its offspring the vacuum, can only exist on condition of not existing at all, and on condition that nothing has ever existed, for if something exists or has existed everything, necessarily, must have existed throughout eternity, and nullity would at no time have been conceivable. If nullity is anywhere it is everywhere. Where would you have this wretched nullity manifest, and where not manifest itself? And when is it to be?

Neither space nor time would be of any use to it, for it can neither occupy the one nor clothe itself with the other. When it seeks to utilise them it becomes that which it utilises. If it were possible, if it ever had been possible, we should not be existing. That we are able to think that it exists proves that it does not exist. We cannot speak of it without giving it an existence which it never has had, that is to say, without annihilating it.

·　　·　　·　　·　　·

Nullity or nothing is the opposite of everything, and just as there is neither nullity nor nothing, there is, strictly speaking, no everything. Everything would be finite; even the "totality of everything" would not be accurate. The number that stands for everything would be unalterable ; but it is impossible that a number can represent a quantity to which, without augmenting it, one can add, over a period of centuries, a billion zeros every second.

.   .   .   .   .

To Fresnel ether was a milieu, the properties of which, inertia and elasticity, were very similar to those of ordinary matter. To Hertz ether possessed neither mass nor elasticity : its condition was defined by the values of electromagnetic fields and by the density of energies. Besides those who hold that it is a fluid subtler than thought there are others who contend that it must be a perfect solid, because experiments in polarisation demonstrate that the vibrations ether transmits are purely transversal, and it is only in a solid object that transversal waves are possible ; some add that ether is incompressible, rigid as a wall of steel and two million times denser than lead.

The most curious thing is that it may well be that all are right, for whether it be fluid, gas or incredibly dense substance we cannot verify it since we are within it, since it is within us, since we are only it in a transitory and particular form, and since everything is in relation to our density. Nevertheless, it seems to me that if we were two million times denser than lead the latter would be nearly two million times lighter than ourselves, and we shall eventually perceive it.

It is not less probable that matter is merely an accident, though infinite, in ether which represents the infinite, and that it is only, quite conceivably, a sort of coagulation of ether.

. . . . .

I have said somewhere in *La Grande Feérie* : " Is it not permissible to suppose that the movement of celestial bodies through the skies is not strictly in their own right, but is due to the ether through which they glide and which is the very soul or substance of the infinite, the only thing that pervades the infinite entirely, and which, like the latter, has never had limits either in space or time?

"This ether in which we are lapped, thanks to which all changes occur, thanks to which all the influences, all the vibrations, all the waves, all the communications, all the manifestations, all the laws of the immeasurable symbiosis—thanks to which all these are transmitted across infinity—this ether has hitherto only been a somewhat controversial hypothesis, a sort of presentiment. To-day it seems to be as necessary as distance or duration. For we can no longer attribute to the absolute vacuum, which being nothing can do nothing, the incessant, the incredible activity of an agency which expends itself prodigally in everything that happens on the bosom of the finite as well as of the infinite, and is perhaps the quivering substance of it all.

"Without it most of the physical and chemical phenomena of former times, and nearly all of the phenomena which have been recently discovered and which have completely upset and abolished the old theories, would be as inexplicable as if one were to misconceive the power of space and time."

I also described it as " the element or the medium of the universal energy " ; it would be more accurate to say that it is energy itself.

.   .   .   .   .

In an unpublished work entitled *Nouvelle théorie mécanique de l'aether* a scientist, M. Henri Joly, asserts and attempts to demonstrate that ether is not a but *the* substance of the universe, that all objects are composed of whirls and twists in that substance, and that all energy is only a movement of ether. The electron deflects the lines of force in its vicinity not by friction—there cannot be any in ether—but through the effect of the vacuum produced between them, a vacuum immediately filled up by the movement of the particles.

What is called a wave, he says, is not something that propagates itself, it is a permanent distortion which accompanies the electron or the photon. The deformation is bound up with them. At the passage of a whirl and in relation to a point in space the ether is deformed, then after the passage reforms itself, it undulates. The electron or the photon presents itself therefore as an undulation; but it is not the wave that guides the photon, it is the photon that guides the wave. The electron is vacuum and the ether is the fullness.

To prove the existence of ether, moreover, M. Joly very rightly observes that light waves are visible in interferences, and that if we rejected ether we cannot understand how there could be a wave in a milieu which did not exist.

The clash of electrons explains both universal gravitation and centrifugal force.

.   .   .   .   .

This theory is somewhat like the one aired towards the end of the eighteenth century by a Swiss physicist named Lesage. According to his theory attraction is due to the clash of corpuscles which he called *ultramondains*, and which moved rapidly in all directions. Corpuscles are indeed found to-day in very highly rarefied gas, where they move in an analogous fashion.

All that is quite possible, but still somewhat uncertain, as indeed is everything concerning electrons—or particles of matter; photons—or particles of light; and the quanta or particles of energy which are invisible entities about which people talk as though they had just been lunching with them. These theories, in which there is more of mathematics and hypotheses than of controlled realities, lack above all else experimental support, and too many questions are left obscure for which an answer must first be found.

.   .   .   .   .

One point on which all those who admit the existence of ether seem to agree is that outside the field of gravitation it is homogeneous and isotrope; that is to say that it presents the same physical properties in every direction. Some add that it behaves, from the kinetic point of view, like a solid and rigid body.

.   .   .   .   .

Since we do not know what gravitation and ether are might we not say, in order to combine and provisionally simplify the two unknown quantities, that gravitation is only a manifestation or a will of ether ? That what we call weight, what we take to be a downward vertical fall, is only movement, that this is beyond everything else the principal phenomenon of the life of the universe or of its substance, ether, the only phenomenon from which all the others derive ?

.   .   .   .   .

If ether does exist it is so powerful, so omnipresent, so universal, so singular and absolute, so infinite, that one may maintain that we and all that exists are but ether, nothing but ether, and that time itself is at bottom nothing but a form, a condensation of ether which represents the great enigma, that is to say that which others call God. This is not to imply that ether knows what it wishes, or that it has a plan or an aim. Why should it have ? Ether is everything and that suffices for it. And thus in the end one catches oneself admitting that this thing of which one is most uncertain is perhaps the only thing that exists.

.   .   .   .   .

Those who maintain that ether does not exist call it space, and the word, representing the same unknown quantity in a less peremptory and presumptuous form, performs the same service.

Space seems to us less enigmatic (one wonders why) than ether, and more obliging than the vacuum or nullity. Besides, we apply these terms more or less at random to all that we do not see, do not hear or feel, and do not touch, though we know quite well that our senses probably perceive not more than a thousandth part of what there is.

.   .   .   .   .

After denying the existence of ether, or considering it to be useless in the restricted principle of Relativity, Einstein admits it into his general theory of Relativity " as a milieu devoid of all mechanical or kinetic attributes, but determining mechanical (and electro-magnetic) phenomena."

" It would determine equally metric rela-tions in the spatio-temporal continuum, the possibilities, for example, of the configuration of solid bodies as well as fields of gravitation ; but we do not know whether it plays an essen-tial part in the formation of the elementary particles of electricity which constitute matter."

This geometric continuum, in four dimen-sions, the ultimate formula to which we come at the end of successive eliminations, is no clearer than all that preceded it and at bottom signifies the same thing, that is to say one does not know what it is.

.   .   .   .   .

Having thus torn it of all its attributes, ether is left finally with but a single mechanical property—immobility. But then it ceases to be of further use. It is admitted that with this sort of ether no one can successfully imagine a mechanical model capable of affording a satisfactory interpretation of the laws of the electro-magnetic field. As Einstein himself puts it : " The laws were clear and simple, the mechanical interpretations clumsy and contradictory."

.   .   .   .   .

What is the action of ether on matter ? Of matter on ether ? It is perhaps in the answer to these two questions that the great secret of the future is concealed, though one can already affirm that this reciprocal action which necessarily must have existed throughout eternity would not produce any new phenomenon. But its scientific proof and application may revolutionise our lives.

.   .   .   .   .

In order to commit themselves as little as possible some scientists see in ether after all a fictitious but indispensable basis for the phenomena of light. Others, despairing of their cause, revert to the hypothesis of a universal substratum—which exacts nothing and is no more than a nickname for the profound and primordial mystery which has been vainly plied with thousands of equations.

The word ether is as the word God: it masks and disguises magniloquently that which we do not know.

.   .   .   .   .

To reconcile the opponents and the partisans of ether Jeans, the renowned eclectic and Relativist, says that the " existence of ether is just as real and as unreal as the existence of the equator, of the north pole, or the meridian of Greenwich. What we have here is a mental creation and not a solid substance."

Let us say no more about it and await what we shall soon be told by the scientists who, feeling urged by time and one knows not what other menace of the future, labour three or four times more fervently, more efficiently than ever.

VI

EINSTEIN

EINSTEIN'S theories have disturbed, much more in appearance than in reality, the Newtonian law of gravitation. The law is so perfect that to-day, after two centuries and a half of experiments and tests, notwithstanding the extraordianry perfection of technique and mathematics, and notwithstanding the invention of space-time geometry, only three disparities have been discovered between observed facts and the Newtonian theory of the great planets.

Of these disparities two, viz. the advance of the perihelion of Mars and the advance of the nucleus of Venus, are too uncertain, too insignificant and too much debated over for us to be obliged to take them into account. There remains the perihelion of Mercury, the *experimentum crucis*, according to Becquerel.

Mercury, sunk into the furnace of the sun, is not easily observable. But it is known that the perihelion of its orbit turns by 572″70 per century. Planetary actions that are calculable do not account for this advance in its entirety. There remains an unexplained advance of 43″49. Jean Chazy considers that the observers and the mathematicians may have started off wrong. It is also assumed that there exists an ultra-Mercurial planet ; that there is a chain of small planets circulating on the hither side of the orbit of Mercury or between the orbits of Mercury and Venus ; or assumed even that the sun is not spherical. *There* is the Newtonian interpretation of the phenomenon.

However that may be, Relativist calculations explain this advance perfectly though some critics of Grossmann in respect of Newcomb's work try to belittle it. It should be pointed out that this is a very remarkable achievement.

What are these calculations based upon ?

It is impossible to explain it without filling many pages with equations. All that can be said is that the calculations start from the famous $ds2$ of Schwarzschild, and those who wish to go deeper into the question may be referred to the specialists, notably to the great work of Jean Chazy on *La Théorie de la Relativité et la Mécanique Céleste.*

Two other and more questionable triumphs of the Relativists are summarised thus by Professor G. Darmois of Nancy University in his *Théorie Einsteinienne de la Gravitation.*

" It gives an entirely new conception of the field of gravitation, leading to the conclusion that gravitation is propagated in waves like light, and with the same velocity, the same wave systems and rays."

Let us note that this has not been proved anywhere, nor is it generally accepted.

" It entails the curvature of light rays in a field of gravitation, such as that of the sun. Experiments stimulated by Einstein's theory confirm the presence of a deviation : only this theory explains it."

(The deviation of a light ray, passing tangentially to the sun, Becquerel tells us, should be double that given by calculations according to the Newtonian law, that is to say 1″74.)

" It should not be said," continues Darmois, " that the experiment would suffice to denote the form of the law ; but the harmony between the figures and the theory is nowhere contradicted by experiment."

" It entails the deviation of spectral lines towards the red in a field of gravitation."

" The deviation which experiment has proved to exist in the solar spectrum is not identical with that anticipated by theory. [A defect in its so-called victory.] But one may consider that the Einstein effect has been proved experimentally by the spectrum of the satellite of Sirius. The disparities which have been clearly established as existing between theory and observation in regard to the solar system are due to other effects. They remain, for the present, unexplained owing to our inadequate knowledge of stellar atmospheres."

.   .   .   .   .

On the other hand, Jean Chazy concl0des his discussion in these words : " The latest observations of Campbell and Trumpler on the eclipse of the sun on 21st September, 1922, do not yet avail to elucidate the question posed. They neither confirm nor shake Einstein's law of deviation. They only seem to indicate—if one might really waive aside all suspicion of systematic error—the existence of deviations in the neighbourhood of the sun, without enabling us to determine its law or its precise extent at the sun's periphery. Moreover, it is possible if not probable even that the deviations result from the superposition of distinct phenomena : refraction in the sun's atmosphere ; the somewhat little-known cosmic refraction of Courvoisier, of the order of $0 \cdot 5$ to some degrees of the sun, of some tenths of a second perhaps to $50''$ of that star ; finally the Einsteinian deviation."

" The question therefore remains in abeyance. Judging by the care with which the observations of Messrs. Campbell and Trumpler were prepared and carried out, one can see that it involves great, perhaps insuperable, difficulties in regard to the *discrimination* of the effect of relativity. In any case, numerous and repeated observations of eclipses added to those already established seem to be necessary and desirable if the problem is to be dealt with on more and more solid and sure foundations."

Let it not be said that this is all hair splitting. Nothing is too small when the greatest law of the universe is concerned ; and it is through the small things that the first glimmer of truth often filters down. I do not believe that the existing glimmer is incontestable. The triumph of Relativity is in short less dazzling than it is proclaimed to be, and I think that Paul Langevin is exaggerating when he says that the " notion of space has enabled us to elucidate the mystery of gravitation."

Note that these interpretations and commentaries emanate not from the adversaries but from the partisans and exegetes of Einstein's theories.

Meanwhile, as Jean Chazy admirably puts it in his conclusions : " Even if the theory of Relativity should one day disappear it would have succeeded at least in as much as it has led

to the discovery of the duration of light rays in the vicinity of the sun.   But this qualified result is not enough ;   we must know if this deviation confirms, quantitatively as well, the law set out in the theory of Relativity."

.   .   .   .   .

The relative triumph of the Relativists, which leaves the capital enigma of gravitation totally in the dark, is obtained only at the cost of two exceedingly rash, not to say irrational, postulates, two at all events highly disputed postulates, viz. that there are no velocities greater than that of light ; and that the universe is finite (also time in consequence ?) but being curved, without limits or edges. In other words, it is a sphere, or rather a hypersphere in four dimensions (the fourth being time) which it is of course impossible to visualise.

In what would this finite universe without limit or edge (that is the latest formula to be adopted), like an orange which is also finite and without edges (but would anyone dare to say that it has no limits ?) stagnate—or drift ?   It is essential that there should be something surrounding it to indicate that it is spherical ; else it would be formless, and we would go back to our classical universe without contours, without edges and limits. But if there should be anything, not itself, around it, it would not be the universe.

According to Einstein, the radius of this universe is 84 million light years. According to de Sitter, however, it is only 2 million light years. According to H. Jeans, it takes 500 million years for light to travel round this spherical universe. The rays do not disappear in a straight line into the infinite, as in the Newtonian or the Euclidean universe, but describe a circle round it, and meet again at the antipodes of the star which discharged them. Becquerel, for his part, imagines an " ultra-macrocosm " which would be one thousand *parsecs*, or 3,200 light year cubes. Upon what are these discordant and fabulous figures based ? Where does science end and phantasy begin ?

·   ·   ·   ·   ·

As concerns many of these cosmic hypotheses (and not particularly these alone) one cannot help thinking that they have at times, in spite of their enormous dimensions, something ingenuous and childish in them, and one wonders how they could be put forward shamelessly before serious-minded people.

.   .   .   .   .

If the hypothesis of the curvature of the universe and the return of light were correct— that is, that the light rays after going round the universe return to their point of departure,—we should be watching the extraordinary phenomena of anti-suns and anti-stars mentioned by Becquerel. At a point in the sky opposite to the real sun we ought to see an anti-sun ; and if there is no absorption of the rays, the phantom sun must be as brilliant as that which illuminates us. It must be the same for stars which can tour round the cosmos two or three times and which must be visible two or three times as ghostly apparitions of the first star. Our firmament would then be nothing but a vault of flames.

By the way, I do not see why the phantom sun should appear necessarily at a point in the sky opposite to the real sun. That is, however, a secondary objection. For the rest, they tell us that if we do not see those illusory stars, it is because there is probably almost total absorption of the rays ; because, again, Einsteinian space is only quasi-spherical and because, finally, light hurls itself against the barrier of time, and " for the observer," says Becquerel, " never would a projectile, never would a ray of light break through that barrier, never would it come back. Yet, if one could measure the velocity of a projectile which is receding, one would find that it increases indefinitely. This gives indeed the finishing touch to our illusion of a universe infinite in space as it is infinite in time."

I confess I am unable to conceive clearly of this barrier of time, and I do not understand how it functions. But since we are dealing here with illusory stars, is it not at least in this case rather the Einsteinian theory that appears illusory, and the universe that asserts itself to be quite incontrovertibly infinite ? Is it not we who are coming out of our bubble universe to re-enter the real one which has no limit in space and time ?

.　　　.　　　.　　　.　　　.

And now we are up against other difficulties. According to some, time travels in a straight line into the infinite. According to others, it is curved like light and space. What sort of time is it which is curved and turns round on itself ? Does it mean the end of eternity ? But what shall we put in the place of eternity ? And further, can time be reversible, time, of which we have always and very rightly believed the essential characteristic to be irreversibility ?

.　　　.　　　.　　　.　　　.

According to Minkowski, one of those who paved the way for Relativity : " Space and time considered apart and in themselves must now recede into the background and only their combination can possess an individuality." *There* are two phantoms about which we know nothing at all, not even whether they exist ; and these two, it seems, by putting together their unknown, their perhaps non-existent quantities, become a reality on which the new theory of the universe is founded. For everything is explained away by the deformation of space-time.

It is a strange and exceedingly prolific union, this—a sort of *Ménage Gigogne* where the children, no sooner than they are released from their apron strings, break into all the sciences and give an answer to every question.

. . . . .

Considering that the universe is finite, there is, says Einstein, no absolute space. And he stretches this theory so as to include the non-existence of absolute time as well—that is to say, the principle that time is not the same for a stationary observer and for an observer in motion.

As it is admirably summarised by Léon Sagnet, the restricted principle of Relativity declares that " it is impossible by any experiment, whether optical, electro-magnetic or of any other kind, to recognise a movement of absolute translation."

From this we proceed to the idea " that the measure of a distance cannot be defined other than through a given lapse of time, and that the lapse of time cannot be measured save at a given point in space. It has hence been concluded that time and space are so closely interwoven that it is impossible to separate them and that the reality is the combination of the two, that is to say space-time." That is the point of departure for generalised Relativity.

The restricted principle of Relativity is based on the Michelson experiments ; generalised Relativity on the four-dimensional universe of Minkowski.

We have already seen, not what one ought to, but what one *can* think of the finite universe, and we will come back to it again. But why deny the existence of absolute time because two observers, of whom one is stationary and the other in motion, do not appreciate time in the same manner. Is not the time of both the observers essentially *human ?* And what is it that, in its own favour, diminishes or anni-hilates absolute time, i.e. eternity, of which the existence is even more indisputable than that of space ?

.   .   .   .   .

" I prefer," says Eddington, on the other hand (as quoted by Becquerel), " to regard matter and energy not as the factors which produce the different degrees of the curvature of space, but as elements of perception of that curvature."

.   .   .   .   .

That is more reasonable and avoids mathematics, but makes of man the sole witness and supreme judge of the universe. We fall back into utter homocentrism.

.   .   .   .   .

We fall into it once more, and indeed become involved in it more deeply, when we are almost constrained to admit as Becquerel admits it, in the wake of Eddington, " that the particles which in the final analysis constitute matter are only a geometrical singularity of the universe, matter ceases to be a primordial entity, mechanical and physical tensors become geometrical

tensors seen under an aspect relative to our
interpretation of Nature, relative to our understanding."

A serious though somewhat muddled admis-
sion, this, which leads us very nearly to the
confession that the "law of gravitation is
completely subjective." That it is entirely "in
the interior of our minds," as Littré said. This
confession has no doubt only been made with
certain reservations, too technical to be gone
into here, but it is none the less real and
worthy of attention.

.   .   .   .   .

Let us note in passing some other astonishing
propositions. "Space empty of matter is not
amorphous," says Becquerel. What does it
mean? If it is empty of matter, it is empty of
everything, and could it be empty, in other
words, nullity, and at the same time not be
amorphous or have contours or limits? "Space-
time is cylindrical," he goes on to assert. It is
possible, but unverifiable. And this: "There
is no essential difference between gravitation

and inertia. The same quality of an object manifests itself according to circumstances either as inertia or as weight ; in precise terms : the force of gravitation is a force of inertia." Of all these statements, the last one seems to be by far the most admissible.

" Gravitation must be propagated action." " To the question ' why does an object, if it is lifted up and released, fall ? ' everyone is tempted to answer : ' Because it is attracted by the earth.' Modern physics must formulate the answer in different terms."

" In the electro-magnetic domain the development of the theory of not-instantaneous but propagated action has led to Maxwell's theory and the restricted principle of Relativity. An analogous idea must be accepted in regard to gravitation. The attractive force of the earth on an object that is falling is, in brief, merely an appearance ; for it is indirectly that the earth acts on the object. In a general way, all matter or all energy determines within its vicinity the properties of space-time, produces a modification of space-time which shows itself to us as that which we call a field of gravitation. The capacity of acting on an object or an electro-magnetic wave belongs to space-time, modified by the proximity of matter or energy ; it is not an effect produced instantaneously and at a distance by an attracting body.

" It is important to note that the deformation of space-time must not be considered to be the cause of gravitation. Between gravitation and the structure of the universe there is no casual link, for they are one and the same. The phenomena of gravitation are simply the mani-festations of the deformation which exists in the presence or the proximity of matter governed by the law discovered by Einstein, and of which the original cause remains a profound mystery.

" It is now certain," adds J. Becquerel, " that a field of gravitation is the manifestation of the non-Euclidean character of the geometric structure of the universe."

    .   .   .   .   .

After a peregrination edged with assertions which, for all their cleverness and audacity, do not even illuminate their own meanderings, we are thus back again at a mystery identical with that of the Newtonian system, with that of all systems and all religions. We have only flourished our words and figures like jingles round the unknowable ; it is something, but we are too much tempted to believe that it is all.

.   .   .   .   .

Let it be added that Becquerel who is, like Chazy, a fervent champion of Einstein's theories, concludes very much as he does, in these words :

" Whatever may be the development of ideas in the future the union of space and time, the inertia and the weight of energy, the law of gravitation, the dynamism of Relativity, the curvature of the universe and the general laws of electro-magnetism are results, nearly all due to Einstein's genius, which will abide with science."

" The theory as it stands may be retouched or completed, above all else in that which concerns cosmological hypotheses and the generalisation of Einstein's theory. But it can be affirmed that a journey backwards to the ideas which are still rooted in some minds is impossible."

If we make a few reservations with regard to the law of gravitation which, until there is a better one, must remain distinctly Newtonian, with regard to the dynamics of Relativity which does not yet exist, and the curvature of the universe which is only admissible in a limited universe, we might share the opinion.

．　　　．　　　．　　　．　　　．

The new theories, moreover, do not agree in essential points. It is said, for example, that the universe is a bubble, or rather, that it is on the surface of a bubble of four dimensions, three of space and one of time; and that between this bubble and other bubble universes which perhaps exist, there is an emptiness traversed by no radiation whatever. It is also said that matter is only manifested through the deformation of space-time in four dimensions; that space is curved owing to the matter it contains; that it would not exist if there were no matter at all, because it is the latter which creates space by displacing itself, whereas others are of the opinion that matter destroys space by occupying it.

There are those who declare that space is an elastic solid, while others maintain that it is incompressible. Some hold that gravitation is caused by the longitudinal waves associated with ordinary light and radiant heat, or that it is due to the clash of electrons. And the most widely read and peremptory writers conclude that it is a geometrical property of the universe —a conclusion which, like the *virtus dormitiva* of opium, in the *Malade imaginaire*, explains everything and suppresses all discussion ; and when Paul Langevin, one of our most eminent scientists, affirms that " we must envisage a geometry determined by the real contents of the universe," we wonder if geometry determined by anything else has ever been imagined. The whole thing is to know what is the real content of the universe ; and I do not believe that even to-day we can flatter ourselves by supposing that we know it.

.   .   .   .   .

In sum, what these writers add to our knowledge of gravitation is highly deceptive. Their most reasonable affirmation, out of all those that are not simply more or less wild hypotheses, is that gravitational forces, like the force of inertia, are determined by the spatio-temporal distribution of the matter in the universe.

The influence of the totality of matter scattered in the cosmos, as explanatory of the phenomena of gravitation and inertia, is an idea, *ad libitum*, masterly or trite. For everything depends evidently on what there is in the totality. However, the idea is unverifiable and leads us nowhere, for we do not yet know what is the totality of this matter.

. . . . .

The new law is above all a triumph for mathematics. It sets up, as Eddington says, a geometrical quantity, curvature, where the Newtonian law had set up a mechanical quantity —force. It would seem to prove that the thought—if thought there be—which presides over the existence of the universe is not the

thought of a mechanician, as in the Newtonian
theory, but of a mathematician or a geometrician

It is not yet known how far the equations o
Einstein and his followers correspond to reality
and it has not been at all demonstrated, as they
affirm, that gravitation is a geometrical property
of the universe. The weak point is that nearly
everything depends on calculations which are
not necessarily founded on observation.

.   .   .   .   .

Let us add that Einstein's cosmology has
been modified by a new cosmology of which de
Sitter of Leyden set forth the rudiments some
fifteen years ago. De Sitter's universe is a
bottom only a different cosmogonic solution o
Einstein's equation. It is an empty universe and
there is no absolute time. It is like an
Einsteinian universe of which the radius varied
with time. All in all, these are mathematical
diversions and conflicting equations the formula
of which are too complicated to be given
here.

This is one of those moments when it is

useful to bear in mind Langevin's words, that science is only somewhat advanced common sense.

.   .   .   .   .

These, then, are the main items in the indictment and the essential arguments for the defence. Those arguments do not conceal the weakness of the system. As for those who indict or attack it, a discussion of them would take us too far into the driest deserts of mathematics. It is, besides, difficult to refute or destroy what has but barely begun to exist. Here are the data before you. I am not competent to be a partisan. And regardless of who wins or loses, let us look only for the truth.

The main point is the impalpable dogma of the absolute constancy of the velocity of light; a constancy which, as Lieutenant-Colonel Corps says, has no other basis than the arbitrary determination of a fictitious local hour. Other points are the curvature of the universe, and of time, and the question of ether, first of all denied, then partially accepted by Einstein, and admitted by the majority of the Relativists.

.   .   .   .   .

To sum up—the problem is not ripe, and there is no reason to be surprised that it is not Newton's theory met at first with the same opposition, and it took more than two hundred and fifty years of observation and verification finally to establish its infallibility. The adumbrations of Relativity go back no more than half a century. It is the part of wisdom, therefore, to wait until it has in turn been submitted to the tests of time and innumerable observations. Only facts which are the voice of Nature or of the universe can disentangle and authenticate the revelations which it claims to have brought with it. Let us recall Newton's wise words : *Quidquid enim ex Phænomenis non deducitur, Hypothesis vocanda est et Hypotheses seu Metaphysicæ, seu Physicæ, seu Qualitatum occultarum, seu Mechanicæ, in Philosophia experimentali locum non habent.*

VII

THE EXPANSION OF THE UNIVERSE

A GROUP of Relativists, with Eddington, the great commentator on Einstein's theory, prominent among them, finding themselves equally constrained in the German mathematician's universe which does not explain, for example, the recession of spiral nebulæ proportionally to the distance; and in de Sitter's universe which explains the phenomenon but is empty: which is inadmissible; are rushing to-day into the expanding universe, the elastic or India rubber universe, of the Abbé Lemaitre. It is no doubt the first step in the somewhat embarrassed return, by a circuitous route, of the prodigal children to the traditional infinite.

.   .   .   .   .

In order not to withhold from it all movement—from this universe of which the magnitude or rather infinitude forbids displacement, this unhappy universe congealed into nullity— the theory accords indefinite expansibility to it. The universe is growing every moment like a soap bubble that is being blown. This theory is founded on a real and quite recently observed phenomenon, the recession or flight into space, towards we know not what far periphery, of the spiral nebulæ which are the remotest celestial bodies and which may be observed through the giant telescope at Mount Wilson.

The spectrographs of these nebulæ show in fact a shift of H and K rays towards the red. This shift, interpreted with the help of the Dopler-Fizeau principle, proves that the spirals recede with a radial velocity which increases in proportion to the distance. " In 1929," says Henri Mineur, the astronomer at the Paris Observatory, in his brochure, *L'universe en expansion*, " Hubble and Humason found that the farthest nebula recedes at a speed of 12427·4 miles a second, that is to say, a fifteenth of the speed of light."

Upon this recession, Friedmann, and later the

Abbé Lemaitre, professor at the university of Louvain, built up their theory of expansion in support of which the Abbé Lemaitre has adduced very recondite equations resting on a hypothesis which is quite defensible, provided we do not drag in the absolute universe but a very small fragment, even if it be extra-galactic, of the visible universe. We cannot in fact imagine the expansion of a universe which has neither outline nor form, nor end, though we can conceive of a small visible universe which recedes into the great All. We shall never, however, be able to see it, either because it would always transcend the range of our telescopes or because it is formed of substances or essences not visible to our eyes.

It is not therefore the universe which expands, but one of its bubbles which swells and gets displaced.

It is some time since Einstein, Heckmann and de Sitter, taking up an idea that Friedmann had had in 1922, admitted that it is not necessary, to prove the Einsteinian equations, to suppose that the universe is spherical. Evidently it is neither spherical nor cubical nor conical. It can have no form since a form always presupposes a contour. But what becomes then of the famous curvature which supports the whole edifice ?

.　　　.　　　.　　　.　　　.

All these hypotheses and all these calculations are so arbitrary that one can make them say exactly the contrary of what they affirm and they would be equally corroborated (it has been proved) if we were to apply them to a universe in process of contraction.

.　　　.　　　.　　　.　　　.

After all, it is quite possible that we are living in a region of the universe in an expansive or centrifugal period, a universe perhaps that is exploding, since the explosion of a universe would last probably for millions of years. It might be maintained with reason that this extension or explosion began only recently, that is to say some thousands of centuries ago. Besides, whether **it be expansion** or contraction, we should know nothing of it, we should perceive nothing of it.

This extra-galactic flight or fall may only be an ascent or a rally round a different centre ; for in a universe that has no landmarks, everything might quite well happen contrary to what we believe. On the other hand, is it not natural to suppose that expansion, if expansion there be, was preceded by a contraction, a coagulation, in which all the stars or all the nebulæ converged towards a centre, clustered together, and formed roughly an approximation to that single and omnipresent block of matter which is the terror and the final destiny of all the spheres ?

Does not all this represent a return to the great Vedic hypotheses which are about a few hundred centuries old and according to which the universe is only a dilatation, an exaltation, an emanation or an expiration of God who perpetuates himself for billions of millenia after which the bubble is deflated or bursts, and there is once more an inhalation, contraction, re-absorbtion or return into God who, for other thousands of millenia, without destroying anything—for everything is indestructible—renders invisible that which had been visible?

And since it is absurd to suppose that witness, human or superhuman, is necessary to what is taking place, the universe would continue to exist latently, as though man were still present. The universe is its own witness—that is all and that is enough.

.   .   .   .   .

In speculation on these matters, the question has been raised whether the universe is forming or decomposing. It is very probable that it is doing both at the same time; that it has been from all time and simultaneously, and that it would always be—in process of formation and in process of decomposition; or rather that formation as well as decomposition, the beginning as well as the end, creation as well as annihilation, are myths existing only in our imaginations.

.   .   .   .   .

In relation to us, our solar system, as we perceive it, is of unshakable stability. Nothing has changed during the historical ages which are, indeed, reduced to no more than some thousands of years—that is to say, to some seconds, if the duration is considered in proportion to the mass. It seems to us prodigiously stable and probably eternal only because it is prodigiously immense ; just as a flower must seem unalterable to the insect which flutters round it.

In reality, everything in the solar system is evolving ceaselessly, and tending apparently to lead the planets back to the central mass. To human eyes all was, all is and all will be collisions, conflagrations, explosions, catastrophic fusions, but in reality they never have been, they never will be disastrous. We forget that nothing can happen to matter to diminish, augment, wound, degrade, destroy or make it suffer. To it, everything is but a superficial transformation, evanescent, and of no moment. Out of the tests it undergoes—which are but stages in its life—it always emerges victorious and happy because it is immortal. A day will perhaps come when we shall cease to believe that the ideal of the universe, or God, is repose.

.   .   .   .   .

Let us not overlook that—though the distribution of stars in space and the arrangement of matter in those stars change ceaselessly—the laws that preside over the changes never vary. What is most astonishing in nature is the superb and implacable logic, the inflexibility, the infallibility, of the rules which we have been able to detect at work.

Whatever may be the perturbations, the cataclysms which overwhelm it, whatever be the terrible and unforeseen circumstances in which it finds itself, every electron, every atom, like every sun in every nebula, knows at every moment what it should do, does it faultlessly, inevitably, cannot do aught else and can never go wrong. At the origin of all the laws which seem to us of a senseless complexity and of which we have barely begun to disentangle some threads, there is always gravitation—the enigmatic, the simple, the irresistible, the ineluctible gravitation.

No law of nature has ever been, ever will be, ever can be transgressed; else nature would cease to be what it is, or rather, would cease to be, altogether. The transgressions which we believe to have discovered only occur in our ignorance.

. . . . .

In man's eyes the visible flows ceaselessly out of the invisible and is only a trivial and purely human incident. In God's eyes it is the entire invisible order which is gushing forth ceaselessly, or rather, is always in ebullition.

. . . . .

Eternal does not mean immutable. Movement is the only life of eternity that our senses enable us to discover. Why should a universe existing from all eternity and for all eternity be less acceptable or more incomprehensible than the endless and originless god extolled by the great religions? Save for the anthropomorphic bias, are they not the same?

. . . . .

The movement the senses reveal to us in the life of eternity is the movement of matter ; and matter, as Joseph le Boucher said, possesses movement as the triangle possesses three angles, by its very constitution.

.   .   .   .   .

Nothing having ever been created, nothing can ever be created. The universe can never be augmented or diminished. What is removed from it would not leave it, what is added would already be within it.

.   .   .   .   .

The expanded universe seems as phantastical as the curved universe. They are both working hypotheses ; but they confuse what happens in the universe with the universe itself. " What real, effective movement can the universe possess ? " asks M. R. Delaunay who, without knowing the Abbé Lemaitre's theory, has come to the same conclusions.

" It can effect but a single movement, that of dilation. The universe can dilate itself since nothing stands in the way of its indefinite expansion. Expanding, the universe still remains the universe. It will remain a unity whatever may be the dimension which the extension may cause it to attain."

How can we speak of dimensions when we are speaking of the universe?

Who says dimensions says limits.

"Since the universe," adds M. Delaunay, "has the possibility of effecting a real movement, one might be assured that it effects it. The universe never seeks to realise the impossible, but it always accomplishes what is possible, all that is possible."

Perhaps; but it does not accomplish what is useless or absurd. In order that it may be able to expand, there must be a free or empty space round it; but that space would already be the universe and it cannot find within or around itself a single spot where it already is not and which it is not. To expand, is it not to displace its surface, occupy a place that it was not occupying before?

The space it conquers is itself under a different name. Space as well as time, matter as well as energy, constitute the universe. It would therefore be more correct to say that, in dilating, it simply modifies the distribution of matter or energy within its totality.

If it could dilate, it should be able to turn round itself. Universal rotation would then be a consequence of the rotation of the universe, just as gravitation would be a consequence or a reproduction of its fall into its own infinity. It is quite possible that we have here both rotation and a falling of the matter found scattered therein. This eternal falling, in the eternal immobility of all things, would then be no more than a rotation in a direction inverse to the first.

．　　　．　　　．　　　．　　　．

But the rotation strictly so called, of the universe strictly so called, is inconceivable. Being everything and everywhere, it cannot turn round in anything not itself and not turning at the same time. It can only rotate in relation to an infinite series of fixed points, and between those points—assuming that it rotates

—it would only be the finite universe which we wish to get rid of.

.   .   .   .   .

The misunderstandings are due to the imprecision of the word universe ; there are as many different universes as there are heads. I create mine as you create yours, but neither of our universes can transcend the limits of our understanding or our imagination ; and no one has constantly present in his mind the image or the notion—which it is of course impossible to picture to oneself—of an absolute universe.

.   .   .   .   .

We assume too readily that the only possible universe is the one we see, as though our eyes were the only testimony to all that exists. We only have confidence in that sense which deceives us more often than the others. Before the invention of the first telescopes, before the latest improvements in them, we did not see a hundred thousandth part of the world we have discovered to-day.

In those telescopes we put the confidence we

have just withdrawn from our eyes, forgetting that they represent but an enlargement of the eyes in which we have ceased to believe. Even when these instruments are ten thousand times more perfect and powerful, it would still be not they but exclusively our eyes which would gaze at infinity. The same remark could be made in regard to the microscope. We can never escape our dependence on our eyes. We imagine presumptuously that the eyes of which our optical instruments disclose the infirmity have alone the capacity to authenticate the existence of a world created, it would seem, in their measure and for their purposes.

Even photography, which we believe to be incorruptibly objective, very probably deceives us. To begin with, the photographic apparatus has only been constructed in imitation of our eyes. Secondly, when we have photographed the firmament, it is still our eyes which examine the plates or the prints and find therein only forms, appearances, light and shade which are of their own world, that is to say, that they rediscover themselves in space ; and the forms and appearances do not perhaps represent more than a hundredth part of what the unknown has imprinted theron. Who could tell us of all that is on the plate, by the side of and encircling what we observe there, and that an eye other than the human eye would see, it may be, with admiration or horror ?

    .   .   .   .

On all sides and in all the sciences, our vision encounters the most damaging disavowals. It does not matter. For us, all that the eye does not see does not exist and is necessarily nullity. If we are not on our guard, this idea, which we know to be fundamentally absurd, would rear its head and falsify our reasonings.

Even a very slight modification of the eyes would suffice to reveal by the side of, or beyond all the stars and all the matter surrounding us, presences and energies as important and quite as real, of which we shall never have the faintest idea.

    .   .   .   .

There is nothing to show that, beyond the sidereal immensity which is enlarged in proportion to the size of the lenses in our telescopes, and which we believe to be infinite, there does not lie another universe that has nothing in common with the one we glimpse, one which is in relation to the infinitude of the cosmoplasm what our solar system is to the extra or meta-galactic worlds.

.   .   .   .   .

Astronomers and astro-physicists declare freely that, thanks to spectroscopy, they are able to assure us that the furthest planets do not contain any substance of which the equivalent may not be found on our earth. They infer from this that the universe has only a deter-minate number of elements. We may grant this, provided that the universe in question is that which through our telescopes we perceive or divine ; provided that beyond that universe there does not exist an infinity of others which have nothing in common with it ; provided, moreover, that we should be unable to see them even if one day in the future they were to come within the range of our telescopes because they are composed of substances either invisible or totally alien to our senses.

However far the telescope at Mount Wilson may penetrate, what it would disclose to us would always be the same island, the same bubble, even when it touches the ultimate spiral nebulæ ; and this island or this bubble of which the surface is millions or billions of light years, is only a point in the ocean of the All.

.   .   .   .   .

VIII

MATHEMATICS

ONE amazing hypothesis is constantly justified by the Relativists with another not
less alarming, while the first is surreptitiously
converted into an established fact to be utilised
in consolidating the house of cards. As
J. Mackaye says in *The Dynamic Universe :*

"Everything is explained by mathematical
fictions called gravitational inertia or curvature
of space-time; purely verbal explanations and
calculations of the great mathematicians which
the mathematicians and those who listen to
them take in good faith to be realities."

For there has been an outburst, since
Einstein's discoveries, of a sort of mathematical
frenzy or bacchanalia which must make us
cautious. There are mathematicians who are

already beginning to feel uneasy. " Some mathematicians," it has been remarked, " consider the reality of things to be a minor issue, and have a tendency to substitute their calculations for the object of their calculations, thus taking the shadow for the substance. What, for example, is the mathematics of wave mechanics, when it is not known what these waves are or how they exist? The formulæ of abstract mathematics lend themselves to every hypothesis. The calculations are correct, but they are not rooted in anything. One only discovers in mathematics what one has put there."

When it is said of a theory or a book that it is " very mathematical," it is supposed that everything has been said about it. And even the specialists accept the formulæ without examining them, confidently and with their eyes shut.

·   ·   ·   ·   ·

Mathematics, said a great algebraist, creates nothing and contents itself with transforming elements given from outside.

It is some years since I said in *La Vie de*

l'Espace : It is evidently within us that everything mathematics leads us to glimpse is to be
found.   It simply expresses what we are no
longer able to say, what we are no longer able
to think.   When we suppose that it takes us
beyond ourselves it only proves that we
transcend ourselves often, without our knowledge ;   and when they conduct us into a
superior space, a space in more than three
dimensions, they affirm that this space exists
really within us, for us, and that it has been
awaiting us since the beginning of the world.

It could therefore become one of the most
curious means of investigation, an unforeseen
interpreter of the latent man, or the sub-
conscious.

I regard it as a sort of mediumistic apparatus.
More or less defensible, it is to-day overrated.
The universe is put into equations, as the history of France was put into madrigals.   One
must really bring oneself from time to time to
define the function of mathematics.   Does it
ever occur to a physician to heal his patients
with the help of algebraic symbols ?

An equation is after all only a train of reasoning extremely and often excessively abridged, and its premisses may be as false, as uncertain, as ill-founded as those of any verbal reasoning. The concrete is pushed aside to make room for the abstract and the formula triumphs over the reality. But a reaction is visible, that is taking us back cautiously to experimental verifications which must always have the last word.

We have, in truth, been too disposed to believe that mathematics is the *ultima ratio* of all we know, and that it is almost childish to pay heed to logic, to intelligence and to the simple and splendid utterances of common sense, all of which have just as much right to be heard as their more or less illegitimate offspring, mathematics. He, Newton, too, was a great mathematician, and if his calculations have won the victory it is because much more directly and modestly than the calculations of our Relativists they were subordinated to reason and experiment.

The practice of the higher mathematics—we

have more than one example of it—does not imply necessarily high intelligence, but only the skill to manipulate symbolic signs that are too often arbitrary and mask doubtful or illusory realities which could be made to say almost anything one wished. They only advance further into the unknown when their ways are prepared for them by the intelligence, in which case they confirm now and then an intuition or presentiment of the intelligence, and bring with them the beginnings of control or certitude.

It is so hard to check some of these extravagant calculations in the void that an eminent mathematician, Bertrand Russell, has said, in a celebrated sally, that mathematics is a science where one never knows what one is talking about and whether what one says is true. And echoing him, Eddington, himself, prince of mathematicians and Einstein's right hand, in a moment of lassitude could not refrain from declaring: " The whole thing is a vicious circle. The law of gravitation is a put-up job."

IX

CONCLUSIONS

LET us summarise in a few lines, completing them at need, not the certainties —there are hardly any—but the main statements we have been enabled to make in the course of this journey through the obscurities of the Great Law.

All bodies attract each other in direct ratio to their mass. Why? Neither the Newtonians nor the Relativists have brought to that question the beginnings of a cogent answer. It is an observed phenomenon, a fact and nothing more.

And in inverse ratio to the square of the distance. Why the square, everywhere and always, without any exception (since, as we have seen, the case of the perihelion of Mercury cannot be held to be an exception)? Why, whatever may be the star, sun, planet, nebula or white dwarf, whatever its chemical composition, its temperature, its density, its remoteness, why never the double, the triple or the cube ? Why should a number multiplied by itself be the prime law and the magic formula of the universe ? The same answer : it is an observed phenomenon, a fact which we are not at liberty to repudiate or recognise. We must bow down to it; it asserts itself with all the weight lent to it by the earth and the stars, though no calculation, no investigation, has shed the least glimmer of light on its origins, its causes and its ends.

.   .   .   .   .

Newtonian physics and mechanics admit that they know nothing as to what gravitational force is in itself, and how it can operate instantaneously across the most fantastic distances. They simply study its effects, and relegate the rest to the rank of unfathomable mysteries, such as life, being, infinity, eternity, time, space and, in general, if you look into the depths of things, nearly all that exists.

The Einsteinians, on the other hand, do not own up to anything, and claim that " the notion of force is a creation of our minds which corresponds to no reality in nature where there are no forces but deformations and movements. It is the acceleration of these movements which causes the phenomena of gravitation."

" If we release," they say, " a mass of one gramme the only reality we observe is that this mass puts itself in an accelerating motion in a vertical line, and what we call its weight is precisely this acceleration : the cause of weight is therefore within the body that falls, and it seems futile to go to the centre of the earth to look for it."

In other words, weight falls because it is heavy, and it is heavy because it falls.

I persist in believing, with more modesty, that the weight of a body is in the force which attracts it, and cannot help observing that the same ounce would weigh only a few drams in the moon of which the mass is twenty-five times less than that of the earth, and about twenty-eight times more in the sun, of which the mass is 324,439 times that of our globe. In other words, a man of 176 pounds would only weigh a little more than 29 pounds in the moon and 4928 pounds in the sun, about as much as a 20 h.p. car, which means that he would be flat as a jelly-fish; that it would need a crane to stand him upright; that he would be incapable of lifting his hand and that his bones would be as brittle as a glass rod.

The cause of this weight that varies—is it in the body that falls or in the mass of the star that tugs at it?

And what is gained by describing as deformation, movement or acceleration, that which is called force? Do not these words hide, under a futile mask, which serves no serious purpose and deceives nobody, the same unknown energy?

.   .   .   .   .

Similarly, because it has been inferred (rightly or wrongly) from a deeper study of electromagnetic phenomena that there is no physical action at a distance, Einstein affirms, as we have seen above, that " it must not be said that the stone is attracted by the earth, but that the earth acts indirectly on the stone. The earth engenders in its vicinity a field of gravitation which acts on the stone and provokes a falling movement." I dare say it does, but is " the field of gravitation " anything more than a verbal explanation ?

Nevertheless, it cannot be disputed that the notion of space-time, which has not been thoroughly elucidated, and which is equivalent to movement, has become indispensable because time and space are really indispensable, and the one cannot exist without the other.

At bottom, the new four dimensional universe is quite simply our old Euclidean universe of which the three classical dimensions, instead of remaining stationary on paper, have started marching in a space without limits which unrolls before them time without end.

.   .   .   .   .

Whereupon it would be asked, what is time —that is, the bit of eternity sliced up by man, and whether instead of being a form of our minds, our minds are not rather a form of time.

The question would be above all to know whether absolute time, that which according to Newton is flowing everywhere and always in the same manner, really exists or is only eternity which does not flow.

The concept of time has attracted the attention of scientists and metaphysicians more recently even than gravitation, which already dates back three or four centuries. Only the poets have up to our own day occupied themselves with it transiently, to deplore its flight and its ravages, without looking into it any farther or any deeper. But this would require another book.

   .   .   .   .   .

The Relativists say that space is curved because bodies gravitate; but they forget that they have curved it on account of gravitation.

   .   .   .   .   .

Is gravitation, in short, a property of matter or ether, the very life of matter, of space or of the universe? Were any of these hypotheses to be proved, it would teach us nothing. Although it is our own substance, matter is more unknown to us than even gravitation as it is infinitely more complex. According to the latest theories, matter is composed of holes in ether. To Lord Kelvin, it is the place, that is to say the aggregate of the points, where ether is animated by eddying movements. To others, it is the place and the points where ether undergoes a very peculiar twist; to Riemann, the spot where ether is constantly being destroyed (but can ether be destroyed ?). Still more recently it has been maintained that ether is non-atomic matter, while the matter we know is atomic, that is to say made up of atoms in motion, as it is also said that this matter is only an appearance of electricity. It is probably true, but then instead of asking, What is matter, we would ask what is electricity ? Whereupon we must ask what is gained by simply shifting the question.

  .   .   .   .   .

All these conjectures, of which it would be easy to lengthen the list, show the prevailing confusion in the new theories. Nearly all of them, however, come back to ether. And we have seen that it has not been proved that ether exists; that it is desirable, not to say indispensable ; that it is not possibly only a creation of our minds and not a real substance. To adduce ether therefore amounts to explaining one unknown quantity which exists by an unknown quantity equally obscure but which perhaps does not exist.

． ． ． ． ．

And now, let us borrow from Einstein himself the final words which should mark the true limits of the controversy :

" Newtonian mechanics and the theory of general Relativity lead to identical conclusions. The concordance between them extends so far that, among such of the consequences of the theory of general Relativity as are capable of experimental proof, very few have yet been found which cannot be deduced from the older physics ; and that notwithstanding the profound difference in the fundamental hypotheses of the two theories."

． ． ． ． ．

Having come to the end of our task, once more, as is always the case when the simplest enigmas of the universe are handled, we find that instead of having given a picture of what is known, truth has constrained us to outline a sketch of what is not known. Let us not be discouraged. From these accumulated negatives will perhaps emerge some day a magnificent positive proof, when everything obscure would become luminous. If we are not yet at liberty to promise, we are at any rate free to hope for it.

.   .   .   .   .

Meantime, look where—after having studied and examined these problems with a method, a logic, a patience, an intelligence which they had not shown before, and with the help of instruments of which until now we have had no idea—look where mankind has got to. Never have they plunged farther into the dark, because never have they searched for the light more avidly. By dint of digging they have only deepened their ignorance. Let us not lament it. It is with gravitation as with all the great problems of the world : the more one studies them the more do they become covered with obscurity ; but an instinct nothing avails to discourage whispers to us that these obscurities are more fertile than the trivial clarities which cradle the slumber of self-complacent ignorance.

In groping through this law, which is the greatest law of earth and heaven, we have once more sought for the will and the purpose of the universe, that is to say for the will and the purpose of God. If this God has not yet drawn aside the infinite that veils His countenance, in proportion as He hides Himself, we discover that He is great. I believe that it is in seeking Him that we praise and worship Him as He would wish to be praised and worshipped ; and that far from blaspheming Him we exalt Him in humbly saying that we do not yet know Him.

THE END